Liquid Lean

Developing Lean Culture in the Process Industries

Liquid Lean

Developing Lean Culture in the Process Industries

Raymond C. Floyd

CRC Press is an imprint of the
Taylor & Francis Group, an **informa** business
A PRODUCTIVITY PRESS BOOK

Productivity Press
Taylor & Francis Group
270 Madison Avenue
New York, NY 10016

Printed in the United States of America on acid-free paper
10 9 8 7 6 5 4

International Standard Book Number: 978-1-4200-8862-5 (Hardback)

Library of Congress Cataloging-in-Publication Data

Floyd, Raymond C.
Liquid lean : developing lean culture in the process industries / Raymond C. Floyd.
p. cm.
Includes index.
ISBN 978-1-4200-8862-5 (hbk. : alk. paper)
1. Manufacturing industries--Management. 2. Chemical industry--Management. 3. Production management--Quality control. 4. Manufacturing processes--Quality control. 5. Industrial efficiency. I. Title.

HD9720.5.F57 2010
658.5'1--dc22 2009044531

Visit the Taylor & Francis Web site at
http://www.taylorandfrancis.com

and the Productivity Press Web site at
http://www.productivitypress.com

I wrote this for my wife, Marsha, and my daughters, Erin and Allison.

They have been the best part of my life.

Contents

Foreword

In business, as in life, having good ideas is only part of the key to success. The ability to communicate those ideas and put them into action is the real test. In a career spanning four decades in both the mechanical and process manufacturing industries, Ray Floyd has repeatedly met and exceeded that challenge.

Ray is a proud practitioner of what is now known as "lean manufacturing." As he describes in these pages, the principles of "lean" are all about engaging frontline workers to help manufacturing plants identify and reduce wasteful practices and to achieve continuous improvement across their operations.

The starting point for all this is something that reflects one of my own core beliefs, based on more than three decades of experience in the oil and gas industry. What Ray and I have both learned is that the vast majority of frontline workers are honest, hardworking individuals who want to deliver top-flight results. The job of management is to give employees the tools and the encouragement to do the right things in the right ways.

The result is a classic "win/win" situation. For employees, the workplace becomes a more engaging, creative environment where they have the opportunity to make a real difference in the reliability and quality of day-to-day operations. For management, there is the revelation of how significantly our plant operations can be improved when we set forth clear strategies and goals and then empower the real experts—our frontline workers—to deliver the goods.

Like Ray's earlier book, *A Culture of Rapid Improvement: Creating and Sustaining an Engaged Workforce,* this new volume is aimed at people who want to lead change in their organization and those who will help and advise the leaders. I firmly believe both books are "must read" material for managers and leaders involved in plant operations of all types.

In *Liquid Lean: Developing Lean Culture in the Process Industries,* Ray draws on his wealth of experience as a senior executive with General Motors, Exxon, and, most recently, Suncor Energy. The book focuses on applying lean manufacturing principles to the chemical and process industries. As Ray points out, by "going lean," these industries stand to benefit even more than mechanical manufacturing operations such as auto plants. That is because, relatively speaking, liquid plants are more capital intensive and any interruption in the industrial process is very costly. This is particularly true of continuous process industries like Suncor's oil sands operation, which runs 24 hours a day, 365 days a year, and relies on a complex series of steps to transform raw resources into market-ready transportation fuels.

As I write this foreword, Suncor is still in the relatively early stages of implementing lean manufacturing principles as part of our renewed corporatewide focus on operational excellence. Yet, even at this early stage, we see tangible results. As Ray documents in this book, lean practices allowed Suncor to reduce costs and increase productivity at precisely the time we most needed to do so—during a period of rapid decline in world oil prices.

True to Ray's time-tested lean manufacturing model, most of the improvements we have made and continue to make flow from the ingenuity and initiative of our frontline workers. I will cite just one example, described in detail in Chapter 9. After our operators were enabled to take a more active role in servicing their equipment, they noticed that pump seals were failing prematurely. The cause was easily corrected, yet mechanics who had previously been assigned to the task from time to time were apparently too focused on replacing the seals to worry about why they were failing. After years of disruptive failures on an important pump, the problem was quickly identified and resolved.

Readers will find in these pages plenty of other accounts of Suncor's journey toward getting leaner—not all of them entirely flattering. Ray is candid about describing times we were convinced one factor was to blame for an operational inefficiency, only to discover the root cause was something else altogether. I have no problem with Ray relating these stories. Like him, I believe the only unforgivable thing about making a mistake is failing to learn from it.

It is important to stress, as Ray does, that making your business leaner is not about making it meaner. After all, one of the worst ways to motivate employees to strive for continual improvement is to use any cost savings that result from their innovations to justify shrinking the workforce. That is why when Suncor initiated a recent major redesign of our organizations for maintenance, support services, engineering, and sustaining projects, we made it clear up-front that the initiative—expected to generate more than $100 million in annual savings—would not result in any job losses.

One of the things I like best about Ray and this book is his ability to address complex issues in concise and easily understood terms. Readers will appreciate how he anchors an entire chapter on the awkwardly named single minute exchange dies (SMED) technology by comparing this lean manufacturing tool to what goes on at your average NASCAR competition. Or how he eases into a discussion of "mistake proofing" your corporation by repeating a baking lesson learned from his mother: Always break your eggs in a separate bowl and examine them before adding them to the mix to avoid one bad egg spoiling the batter.

As a business leader, I know that it is impossible to overstate the importance of operational excellence. This creates the value on which we build our corporate reputation. For our investors—and for all stakeholders—how we perform today determines their level of confidence in how we will perform tomorrow.

What Ray provides in this book is a road map for achieving excellence across plant operations that demonstrably generates tangible results in as little as 6 months and can lead to transformative change, even in a large organization, in the span of a few years.

Reaching out to engage an entire organization in a system of continuous improvement is a very appealing concept and I have witnessed the excitement in the field that comes from pursuing this kind of positive change. I would urge all manufacturing managers and leaders to read this volume carefully and to take Ray's lessons to heart. They will be doing their organizations, their shareholders, and their employees a great service.

Richard L. George
President and CEO
Suncor Energy, Inc.

Acknowledgments

Full credit for the ideas and experiences described in this book belongs to my colleagues at General Motors, Exxon Mobil, and Suncor Energy. Without them, I would have had nothing to report. Several among them deserve special mention. Don Powell, John Webb, John Laibe, and Gene McBrayer of Exxon were inspirational and empowering leaders during the early days and throughout the critical development of this material for application in the liquid industry. Kirk Bailey, Steve Williams, and Rick George of Suncor enabled the most recent work, where we quickly deployed 20 years of Exxon experience into a new business and a new country, demonstrating both the aggregating impact and the portability of the concepts. My colleagues on the Suncor Oil Sands Management Committee have made amazing contributions—individually and collectively—to this work. These men and women all made very special contributions to me and to the resulting experiences described here. I sincerely thank each of them for his or her help.

Michael Sinocchi of Productivity Press deserves a lot of credit for demonstrating remarkable patience when I advised him that I was "unretiring" to join Suncor. The result of this was that the manuscript was a year late and completely rewritten.

Of course, the foundation for everything that I do is my wife, Marsha. Whenever I am uncertain about the right thing to do, I just watch Marsha. Somehow, she always knows.

Raymond C. Floyd

1

Business Results in Process Industries

INTRODUCTION

In 1991, the Exxon business that I had the great good fortune to lead received the Shingo Prize, described by *BusinessWeek* magazine as "the Nobel Prize for manufacturing." Although the Shingo Prize was given "for manufacturing excellence," the criteria were then, and still are, based largely on the work of the prize's namesake, Shigeo Shingo, one of the primary architects of an innovative system of manufacturing variously known as the Toyota Production System or Just in Time and known today as Lean Manufacturing.

Despite the obvious successes of this technology, Western manufacturers have generally been slow to adopt lean practices and put them to good use. Nonetheless, based on the early teaching and examples provided by Toyota, it has long been clear that Japanese practices of lean manufacturing are applicable to the Western automotive industry; indeed, they can be used successfully in most discrete manufacturing. There have even been some well-documented successes in adapting lean principles in other activities that look and feel somewhat like discrete manufacturing, such as the administrative processing of documents.

However, in 1991 when Exxon Chemical appeared, as if from out of nowhere, and was designated as one of the best practitioners of lean technologies, it created quite a stir. Exxon Chemical is a liquid industry and the Shingo Prize designation was based on manufacturing practices that we had developed for a synthetic rubber product that is produced in a cryogenic state through a reactive chemical process. This was *serious* process industry manufacturing. Although that experience was nearly 20 years ago and despite the fact that Exxon Mobil has continued to use and benefit from these practices, there is still surprisingly little use of lean manufacturing techniques in the chemical and process industries.

That is a real shame. The lean tools and concepts fit the process industries wonderfully well and should be a valued part of every process manufacturer's portfolio of operating practices. I regularly describe lean manufacturing practice as an attempt to make the flow of product through discrete manufacturing more like the flow of materials through a continuous process plant. With that concept as a mental model of lean, it is clear to me that the technologies that enable discrete manufacturing to operate as a process plant operates ought to have great value for enabling a process plant to operate even better. As you will see, they certainly do.

HOW THIS BOOK IS ORGANIZED: SHINGO PRIZE CRITERIA

After receiving the Shingo Prize, I wanted to share the lean experience with others, so when the board of trustees for the Shingo Prize invited me to join, I did. I also volunteered to be team leader for the prize's field assessment process for 2 years. Based on this history, it is natural that I have structured my description of *lean* around the criteria of the Shingo Prize, which is organized essentially into four major sections:

- business results
- consistent lean enterprise culture
- continuous process improvement
- cultural enablers

This book will follow the Shingo Prize format for presenting the lessons of lean as they apply within the liquid industries. We begin in this chapter with a description of ***business results*** in process industries where I have led the lean implementation. Chapters 2 and 3 introduce *lean enterprise thinking* and *policy deployment*, the two principal elements of a ***consistent lean enterprise culture.*** Chapters 4 through 9 provide detailed theory, practice, and examples of ***continuous process improvement*** using lean methods in the liquid industries. Chapters 10 and 11 describe *leadership and ethics* and *people development*. These are the principal ***cultural enablers*** of a lean enterprise. Chapter 12 provides a detailed description of how to lead lean implementation during the first 6-month period, along with benchmarks to be achieved during the first 2 years of implementation.

Note: Process industry examples are used throughout the book, but the differences between liquid manufacturing and mechanical manufacturing will be more apparent in the six chapters on continuous improvement than in the chapters on lean theory or culture.

BUSINESS RESULTS: IMPROVE PERFORMANCE WITH LEAN

All of the Shingo criteria have evolved with time as lean manufacturing grew and expanded into new areas such as process manufacturing and administrative operations and as the Shingo board became more adept at assessing successful lean practice. The criteria for assessing the overall business performance of a lean manufacturer were not part of the original criteria, which assessed only the practice of lean. The board added criteria for assessing business results because its members wanted to ensure that they were recognizing businesses that experienced the *substance* of lean and not simply the *form*. As we will discuss often throughout this book, the confusion between form and substance is one of the most serious errors to be avoided in lean practice.

Although it is easy to recognize a business that has adopted the form of lean by the way that it appears and the way that it operates with limited resources, that is not where the true value of lean lies. The *value of lean is in the substance of improved performance* that makes the lean form of operation more effective than traditional manufacturing practice. Although adopting lean usually results in strikingly improved performance, sometimes organizations adopt only the form of lean and not the substance. In those situations, it is quite possible to use lean concepts improperly and to make a business worse rather than better. As you learn about lean and especially as you put lean to use in your business, it is important routinely to ensure that you are receiving the benefits that you anticipated and not merely practicing poor manufacturing in a new way.

Throughout the book, there will be a great many very specific examples of the use of lean technology and the resulting benefits at the field level.

In those examples, you will learn the details of lean practice. Before we get to the details, we will look at four high-level examples of business results that describe the performance impact of lean operation on an entire enterprise. Then we can begin to discuss how you might get similar benefits for yourself.

BEAT THE COMPETITION WITH VERY FLEXIBLE MANUFACTURING

In Exxon's synthetic rubber business, one of the primary raw materials of the reaction is isobutylene. You may recall that on January 17, 1991, the United States and 12 allied nations launched Operation Desert Storm to turn back the Iraqi army that had invaded Kuwait and was apparently threatening Saudi Arabia. Because Iraq, Kuwait, and Saudi Arabia are all huge exporters of crude oil, this first Gulf War substantially disrupted supplies to petroleum markets globally and had a tremendous impact on petroleum prices, particularly on the price of some materials derived from petroleum, such as isobutylene. To illustrate the effect of lean operations on our synthetic rubber business, let us examine how lean practices allowed us to respond to this series of events.

Because fuel demand is relatively inflexible over short periods as compared to the demand for chemicals, the crude oil that was available during the war, especially during the early weeks of the war, was used disproportionately to produce fuel. As a result, the chemical feedstock derived from petroleum refining available for other products became even scarcer than the reduced availability of crude oil at the time would have implied. That effect applied to Exxon's synthetic rubber business; thus, the price that we paid for isobutylene increased from less than $300 per metric ton to nearly $1,400 per ton in a period of just a few weeks. Faced with a cost increase of that magnitude, most suppliers—but not Exxon—in this part of the worldwide chemical industry declared force majeure and canceled their contracts until customers agreed to substantially higher prices.

Because we were then in the early stages of transforming Exxon's synthetic rubber operations from traditional manufacturing to lean, we had another option available to us. At the time, Exxon owned four butyl polymer plants operating in Europe and North America as well as two joint-venture butyl plants in Japan. Our management team decided to continue operations but to shut down approximately 80% of the reactor capacity in each of our wholly owned plants. Most importantly, however, we decided that we would honor our customer contracts.

Note: During this same period, our two plants in Japan, operating as Japan Butyl Company, became the first liquid industry plants to receive the TPM Prize for Total Productive Maintenance from the Japan Institute of Plant Maintenance.

In order to fulfill most customer orders during that period, we shipped product from our existing inventory of finished goods. That inventory had been produced in accordance with our pre-lean standards of traditional manufacturing practice. Only when and only as needed did we use the limited number of operating reactors to produce new product. In that way, we used the then-current high-cost raw material only in limited quantities as needed to balance our existing inventory to customer demand.

It may sound as if our strategy succeeded because we had a relatively large preexisting inventory. Possessing traditional inventory levels at that moment was certainly advantageous, but what made this strategy successful was that, with our new lean capabilities, we now could operate with such flexibility that *we no longer needed to replenish our traditional inventory.* As we made new product to balance our inventory to market demand, we sold 100% of that product at the time it was made. Every one of our competitors also had a traditional inventory supply at the beginning of the war, but only Exxon was able to deploy its inventory in this way. Because inflexible operations generally required our competitors to replenish inventory with high-cost product, they not only experienced the immediate impact more severely than we did, but also continued to experience high costs for a long time after the war as they withdrew this material from inventory.

Because of this strategy, during early 1991, new production, made at the then-current higher price, constituted less than a quarter of our total shipments. With this capability in place, we adjusted customer prices, but only as required to cover the new production as a proportion of our total shipments. We ultimately increased our prices, but our price increase was far less than that proposed by others. Although the market finally priced this product near our price, it was clear to our customers that the relief they enjoyed resulted principally from the fact that Exxon was attempting to honor our existing contracts.

Key idea: Developing new capabilities for your business often enables you to benefit from unanticipated opportunities that arise from changed or previously unrecognized circumstances. An important element of business success is having the capability to take advantage of those disruptive opportunities as they occur.

The lean impact on our business that resulted from exercising our new capability during that period was truly amazing. First, because we held prices down in the market, Exxon's synthetic rubber business operated profitably during a period when our competition lost substantial amounts of money (the price increases that our competitors obtained were often not sufficient to cover their higher costs of manufacturing). As a result, we emerged as an ongoing healthy business while others faced a long period of financial recovery.

Further, because our lean capability enabled us to behave toward our customers very differently than other suppliers did, beginning immediately with the competition's first declaration of force majeure and enduring for a long time into the future, we became the market's preferred supplier. During and after the war, many customers offered to give us 100% of their business.

More than 10 years after I left that assignment to lead other Exxon businesses, I continued to receive regular reports that our customers favorably remembered what we did for them during that crisis. At a time when our entire industry was in turmoil and other suppliers were canceling contracts and imposing huge price increases, customers remember two things about Exxon Chemical: We were honest enough that we honored our contracts and good enough that we were able to do that in a businesslike manner. That reputation has stood us in good stead for many years. Imagine that. Because we adopted a new and powerful manufacturing strategy, we operated much more profitably than our competition and our customers loved us for it.

Lean operations made all of that possible. Remember that part of the title of this section is "very flexible manufacturing." We were certainly fortunate that at the time we needed inventory, we were in the initial phase of our lean transition and still possessed traditional inventory levels. However, the attribute that made this strategy work was not the existence of inventory. With any amount of inventory or even no inventory at all, we would have operated during this period much less expensively and more effectively than our competition. The attribute that made us startlingly successful was the ability to operate very flexibly. We produced no high-cost product in excess of the absolute minimum needed to balance our inventory to our customers' immediate demand. No high-cost material was produced into inventory, and we were able to do that effectively and efficiently with no real losses in either our limited capacity or our expensive raw materials.

Key idea: We know that their (and our) customers told our competitors what we were doing. The simple reason that we were successful and they were not is that *our lean capabilities allowed us to make this operational strategy function* and the competitors could not do it using traditional methods.

We will discuss the benefits of using lean tools and practices to improve the flexibility of process operations in more detail throughout this book, but *the first lesson of lean manufacturing in the process industries* is that inventory reduction is rarely the goal. *The goal is to become an excellent manufacturer.* Reduced inventory is a likely outcome of the new capabilities, but *the strategic focus is on obtaining the new operational capability* and not on reducing the old inventory.

IMPROVE PERFORMANCE WITH LEAN AND AN ENGAGED WORKFORCE

After leaving the butyl polymers business, I became site manager of Exxon Chemical's massive Baytown chemical plant. Baytown is among the largest petrochemical complexes in the world. Although Gene McBrayer, Exxon Chemical's president at that time, had described Baytown when I arrived as "a troubled plant," its performance, although not meeting Exxon's high expectations, was probably near average among all chemical plants in the industry.

Two years later, *IndustryWeek* magazine designated the Baytown chemical plant as one of "America's ten best plants." In a relatively short period, we had progressed from being considered a troubled plant within our own company to being considered one of the best plants among all companies and all industries on the continent. During those first 2 years and for the following 4 years, Baytown produced compounding annual improvement in productivity (output/person) that exceeded 16% each year.

When you do the math, you realize that over a 6-year period, the Baytown team more than doubled the productivity of one of the world's largest and most complex chemical operations. In addition, we positively improved capacity, quality, delivery, service, safety, environment, community relations, employee satisfaction, and, of course, profitability, which more than doubled during that period. Literally, every single aspect of the business experienced a dramatic improvement.

The success was a result of lean manufacturing combined with employee engagement. The Baytown experience has received a great deal of recognition for our success in engaging employees; however, we not only asked people to engage with the business in a new way and to perform better than they previously had performed, but also, through lean, gave them meaningful new capabilities that enabled them to understand objectively *how* they could perform better. Simply asking people to do better rarely results in sustainable improvement, but delivering powerful new capabilities to them as you make such a request can have that result. Lean manufacturing and a culture of engaged people are often closely associated. We will describe the most important elements for creating an engaged workforce in Chapter 10 as part of the discussion of the cultural enablers of lean.

Exxon Baytown did indeed substantially improve employee engagement. During this period, engagement in the improvement process increased to more than 40 implemented improvements per person each year. Compared to the North American and European average for all industry of 0.028 improvements per person each year,[1] Baytown people had engaged with the business and were helping us improve at a pace of autonomous action that was more than 1,000 times the average for Western industry. Engaging every individual so that each person makes his or her personal best contribution to the business is the only way to make so much improvement so fast in so many different areas. During this period, Baytown also received the Andersen Consulting Award for "excellence in managing the human side of change."

Thus, the improvement was not just in productivity; rather, it was in all aspects of the business. In 1995, *Maintenance Technology* magazine designated Baytown as the "best maintenance organization in large industry." For a capital-intensive industry such as chemical manufacturing, where business performance lives and dies with the successful operation of our equipment, that was a deeply satisfying achievement.

There are, of course, competitive implications when one of the largest operations in a global industry doubles its productivity. Although we were always very careful to protect the employment of Exxon employees, we achieved some of our productivity improvement by discontinuing use of some contractors to reduce our total workforce. Of most importance to our competitors, we obtained more than half of our productivity improvement by greatly increasing the capacity and capability of our operation and filling that capacity with new production volume.

At the end of this 6-year period, the Exxon Baytown chemical plant was producing 60% more product than at the beginning of the period. In an industry that normally grows (including capital investment) at a sustained pace of about 3% each year, this organic increase in Exxon's production resulted in taking a lot of sales away from a lot of competitors. Further, by successfully improving the quality and consistency of our production, we substantially improved our product mix by moving upscale to higher value products within the original product families.

There were also traditional business implications for our shareholders. We achieved our remarkable growth in capacity and capability through improved operations rather than capital spending. In fact, we achieved that 60% growth in capacity with new capital less than depreciation or, as we liked to say, with "negative net investment." It should be clear that with efficiency doubled, capacity increased by 60%, a portfolio of higher value products, and a nice reduction in net investment, our profit and return on capital employed improved far beyond any prior expectations.

Key idea: *The second lesson of lean manufacturing in the process industries is that many of the enabling technologies that make lean possible are tools that are especially useful to frontline workers and middle managers.* These capabilities allow them *autonomously to identify and implement many small event improvements* that engineers and managers might never recognize and probably would not have time to address if they did recognize them.[2]

Improvement at a world-class pace requires that frontline people operate the business and conduct small event improvement at the same time as engineers and managers continue to implement strategic changes and conduct big event improvement. When frontline people engage in autonomous operation and improvement, which enables engineers and managers to concentrate on more and better big event projects, genuine synergy exists. The extra boost to performance provided by that synergy is always a necessary element in achieving world-class performance.

There never is a question of "chicken or egg" sequence in achieving this synergy. World-class performance only and always occurs as the result of engaging the frontline people. The capability of engineers and managers to create projects for improvement and growth is a fundamental element of routine operations for all process industries. The impressive pace of world-class improvement always begins when a plant adds to that fundamental capability the new source of enhanced performance derived from front-line people autonomously operating and improving the detailed activities of routine manufacturing.

Initially, big event improvement will continue as it did previously and the small event improvement will simply become an added source of progress. As small event improvement matures and as strategic alignment increases, the synergistic relationship between the big events and the small events will become apparent, and you can begin to think about achieving world-class performance.

Note: Frequently, the terms *world class* and *world scale* are used without definition to describe manufacturing operations, including operations that are objectively neither. The definition used in this book is that a world-scale plant is one that has capacity within 10% of the largest plant of its type in the world. Similarly, a world-class plant is one that performs within 10% of the best plant of its type in the world.

GAIN FIRST MOVER ADVANTAGE

Just before joining Exxon Chemical, I led worldwide manufacturing for Gilbarco, Inc., an Exxon-owned business that manufactured equipment in five plants around the world for retail gasoline marketing and other service station needs. The chances are good that when you fill your car's tank, you are holding a Gilbarco product in your hand. (Exxon later sold Gilbarco, which is now a part of Danaher Corporation.)

In 1985, we introduced lean at Gilbarco; we were the first company in our industry to make that move. The result was dramatic. After experiencing the initial performance improvement and recognizing how much further improvement was possible, Gilbarco management created a joint manufacturing and marketing strategy that assumed that, rapidly and continuously, we would become better and less expensive than our competition. The strategy was simple: Marketing would maintain the unit dollar margin on each product sold. As manufacturing achieved further improvements, Gilbarco would pass those cost savings on to its customers.

In addition, as operational performance improved, the customers benefited from improvements in quality, delivery, and service. Moreover, because of improved operating performance, capability, and flexibility, we began to introduce innovative new products at a faster pace than was possible before. It did not take long before our customers responded favorably to this strategy and our share of the market began to grow quite rapidly. Within 2 years, our global market position had more than doubled.

From a performance perspective, during the 2 years that I was with Gilbarco, we doubled our manufacturing output and productivity, with accompanying increases in quality, delivery, and service. During the next 2 years, Gilbarco management doubled productivity again. All told, *productivity quadrupled during a 4-year period.*

Later, *IndustryWeek* magazine designated Gilbarco as one of "America's ten best plants." This is the same recognition that the Baytown chemical plant received. Because Gilbarco management was kind enough to recognize that I was "chief architect of the changes and lead manager during the implementation,"[3] it is often said that I am the only person to lead businesses in both discrete manufacturing and liquid manufacturing to receive that *IndustryWeek* recognition.

Similarly to Exxon's isobutylene polymer business, Gilbarco's competition soon discovered what we were doing. One competitor even took the unusual step of hiring one of our senior manufacturing managers. However, without a comprehensive lean effort, the competition simply could not operate in a way that allowed them to match our performance.

Among the primary U.S. competitors, one did follow us as we reduced price, but with no improvement in its operating capabilities, it was not able to match us in quality, service, delivery, or the introduction of new products. The result, therefore, was poor. Despite similar pricing, they lost sales volume to us because of our improvements in other areas. More importantly, because the company did not have operating performance to support its reduced prices, it quickly became unprofitable, with the result that its product quality, service and delivery, and ability to introduce new products worsened over time. At a product sales price that made Gilbarco very happy, indeed, the competitor was not able to cover its costs. Another principal U.S. competitor refused to follow our pricing lead, and we rapidly grew our business at its expense.

Note: Just before Gilbarco adopted this strategy, a leading Japanese manufacturer of the same equipment opened a U.S. factory, intending to enter the Western market. Because we quickly became so successful, the Japanese manufacturer closed its U.S. plant and returned to Japan within 5 years, without having made a significant impact on the market. I mention this because lean practice is closely linked to Japanese manufacturing. Although lean certainly originated in Japan, it is also certain that Western manufacturers can implement it with equal success. In fact, because Exxon was among the very first liquid manufacturers in the world to adopt lean, when I later became Exxon Mobil Chemical's global manager of manufacturing services, on several occasions I found myself teaching "Japanese manufacturing" to Japanese leaders working in Japan.

Although Gilbarco was an Exxon-owned business, it is obviously not a process industry. However, it does illustrate an important point that is fully applicable in the process industries. Gilbarco's competitors ultimately recovered and began competing again, as did Exxon's competitors in the synthetic rubber business. In both businesses, however, there was a significant period during which the competition was only surviving, rather than truly competing. Gilbarco became the first mover in its industry and profited greatly from that advantage for many years. This is an important lesson because, in many segments of the process industries today, there has not yet been a successful first mover.

Today, about 50% of all American industry has adopted lean—at least to some extent—but the process industries are far underrepresented in that group: Less than 10% of process industry operations employ lean practice. However, recognition of the value of lean for our industry is growing. Certainly in the near future, there will be a lean manufacturer among your process industry competitors; it might as well be you.

Key idea: One critical lesson common to each of these examples is that within 2 years of commencing the transformation, we had implemented lean manufacturing in a way that dramatically changed our performance, improved our profitability, and gave us a commanding performance advantage over our competitors. *Improvement did not suddenly appear all at once at the end of the second year. We made real, measurable, and ratable progress during the first 6 months of implementation and during each 6-month period thereafter.* Lean implementation does not need to be slow and *the ultimate advantage always goes to the fastest.*

ACHIEVE PROMPT IMPROVEMENT

After retiring from Exxon Mobil, I joined Suncor Energy as senior vice president. That was about 1 year ago as I write this. Suncor (now merged with Petro-Canada) is Canada's largest energy business and is generally regarded as the first and still the principal developer of Canada's oil sand resources. Although Suncor is still in the early days of adopting lean manufacturing, it is not too early to report that prompt improvement is possible. In July 2008, we began to practice the technologies described in this book and the results are already evident.

Prior to commencing this effort, Suncor presented a great opportunity for improvement. Full-year production in 2007 was less than in 2006, and, in the first half of 2008, production results were less again. The decline in production during those years was masked by the rapid increase in the price of crude oil.

However, similarly to the experience in Exxon's butyl polymer business, the world changed. In the last half of 2008, the price of crude oil declined from more than $150 per barrel to less than $40 per barrel. Fortunately, that decline coincided with our conversion to lean manufacturing. During the first 6 months of our lean effort, our average output for the full 6-month period was substantially greater than the average production had been during the prior 6 months. During the second 6-month period of our lean effort, we achieved record production from our existing assets. The Suncor team has stopped the decline and turned the business around; we have begun to improve at a good pace. Suncor achieved this result with no changes in facilities. We simply took that which we had and began to practice lean manufacturing.

It is important to note that, because we were able to increase production substantially from our existing facilities, we became much more productive at the same time that the price of oil declined on the world market. Once again, exactly as we had experienced in butyl polymers, the world changed and our lean capabilities enabled us to prosper at a time when our competitors were all struggling.

Key idea: There are two important lessons here: First, lean is indeed a very powerful capability that *can be rapidly deployed* to the great benefit of your business. Second, *lean is appropriate for businesses of all sizes and histories.*

ALL COMPANIES CAN BENEFIT FROM LEAN, BUT NOT ALL DO

Certainly, Exxon has vast resources and capabilities, but that also implies that the company has a very big enterprise to transform. In my experience, most businesses have a capacity for improvement that is approximately proportional to their size and resources. Exxon used its resources to proceed quickly, as has Suncor. Further, lean is appropriate to any initial state of the business. Exxon butyl polymers and Exxon Baytown were both average businesses that quickly became great businesses. Gilbarco and Suncor were businesses that had great opportunity to do better and, indeed, they quickly did become much better.

"These results are not typical." Diet plans offer this warning in their ads using celebrities who have experienced remarkable weight loss by following their programs. The diet plans are the same for everyone, but some people experience great success. Others have only moderate success and still others have no success at all.

Exactly the same thing happens when companies adopt lean manufacturing. Some businesses promptly experience great success and others spend long years and many dollars to achieve very little success. Some businesses abandon lean without achieving anything at all. In extreme cases, some businesses adopt the *form* of lean *without the substance* in a way that causes their performance to become worse rather than better. Although lean is a great tool for manufacturing improvement, like all tools it needs to be used well. The key issue for businesses of any size or performance is to deploy lean quickly to improve at a good pace. Lean practices coupled with a culture of engaged people can provide the focus and the success that you want.

Disruptive Changes

Each of these examples of lean business results represents what economists call a disruptive change. In Exxon's synthetic rubber business, the price of our raw material suddenly increased by a multiple of five. In Suncor's business, the price of crude oil suddenly decreased by a factor of three. In such situations, the disruptive change was forced upon the business by a change in the world in which the companies operated. In both situations, operating very well allowed us to succeed while our competitors struggled. Sometimes the world imposes a powerful change on your business, and you must be able to react with an equally powerful response.

Of even greater interest for most businesses are Exxon Baytown and Gilbarco, where the disruptive change was caused by the sudden emergence of a competitor with unmatchable capabilities. In those instances, we did not experience the disruptive change; we were the source of the disruptive change that our competition experienced.

Each transformation took just 2 years and, in all four situations, new people and new capital were not needed. We took what we had, we made it better, and we did it promptly. That is the essence of surviving an unforeseen disruptive event and imposing disruptive change on your competition. A similar result is certainly possible for your business. In later chapters, you will see exactly how that performance was achieved and how your company can achieve a similar result.

In every case where world-class performance is achieved, it is clear that the leaders of the business had three important characteristics:

1. They had a *clear strategy* that defined with real precision the business performance that they intended to achieve as well as *when* and *how* they expected to get those results.
2. They had well-conducted *initiatives to engage all employees* to pursue those results with autonomous "best efforts" performance from each individual.[4]
3. They adopted *new technical capabilities* such as lean that enabled them to practice manufacturing in a substantially improved way.

Key idea: When I am asked, as I often am, on which of these three initiatives—strategy, engagement, and lean—a business can focus to get the best results, I always answer that a business needs to focus on all three at the same time. Without all three elements of ignition—heat, fuel, and oxygen—you will never create fire, and without all three elements of operational success—strategy, an engaged workforce, and improvement technologies such as lean—you will never create world-class operating performance. The lack of all three elements is the most common reason that many companies adopt lean with only limited success. Many leaders focus only on the new technology and do not simultaneously adopt the other two essential elements of success. As a result, they have a great tool, but do not know what to do with it and do not have enough people to help them use it.

WHY THE PROCESS INDUSTRY NEEDS ITS OWN VERSION OF LEAN

Assuming that these examples have convinced you that adopting lean is a good idea for those of us in the process industries, let us stop for just a minute and explore what is meant by "process industries" and why these industries need their special version of lean manufacturing.

Significant structural differences between the process industries and mechanical manufacturing substantially affect our ability to adopt lean practices in precisely the same way as our colleagues have done in mechanical manufacturing. It is valuable to understand these differences as we begin the discussion. There are three critical distinctions between process manufacturing and mechanical manufacturing:

1. In process manufacturing, the raw material experiences a *transformational* change as it becomes a product as opposed to a *reconfigurational* change.
2. In process manufacturing, the manner of transforming raw materials into products is often *indirect* as opposed to the *direct changes* that occur in mechanical manufacturing.
3. In process manufacturing, the transformation of raw material is frequently *dependent* on time, but mechanical manufacturing is *independent* of time.

In the nature of process industry operations, each of these distinctions makes a very real difference in the way that we need to apply the theories and tools of lean practice in our plants when compared with mechanical manufacturing. I will discuss briefly here the differences from mechanical manufacturing, and then, throughout the book, I will describe the challenges and benefits they offer for the adoption of lean in a liquid process.

Transforming the Raw Material

The first distinction between mechanical manufacturing and process manufacturing is the way in which the raw material is changed as it proceeds through manufacturing. In mechanical manufacturing, the raw material is "reconfigured" but the material itself remains inherently the same. For example, when an engine block has holes drilled into the casting, the casting is reshaped, but the material itself is essentially unchanged. The same is true when copper tubing is drawn to size, when a chocolate bar is wrapped, or when a lawn mower is assembled. The raw material is reconfigured to become a finished product, but the material always remains the same material throughout the manufacturing activity.

In process manufacturing, the raw material undergoes an inherent "transformation" during manufacturing when it becomes an identifiably different material as a finished product. This is a very different manufacturing outcome, achieved by a very different approach to manufacturing. Those differences necessitate a notably different practice of the lean theory and tools. The following examples illustrate how this concept of material transformation applies in different parts of the process industries:

- *Metallurgy:* In metallurgical processing, heat-treating a block of steel will transform the crystalline structure of the metal so that it is either harder or softer or has less internal stress than the original block.
- *Foods:* In baking, a high-density liquid (cake batter) is transformed into a low-density solid (cake).
- *Reactive chemistry:* In a polymer plant, individual molecules of monomolecular gas (e.g., ethylene) combine to form a macromolecular solid (e.g., polyethylene). Alternatively, in a steel plant, coal, iron, and other chemicals combine to create a wholly new material.
- *Separation chemistry:* In a petroleum refinery, a liquid stream comprising thoroughly commingled molecules of many different materials is transformed into several streams, each with a relatively pure

composition of a single type of molecule. The chemical properties of the several new streams are individually very different from the properties of the original commingled stream.

Thus, although process manufacturing is normally associated with liquids or gasses, the differences between mechanical manufacturing and process manufacturing are more fundamental than just the initial state of the material. In fact, heat-treating a block of steel is an example of process manufacturing, and filling quart bottles with motor oil is an example of mechanical manufacturing.

The difference in the nature of the changes undergone by the material in process manufacturing leads directly to important differences in the manufacturing activities that produce those changes, which leads to differences in the way that we practice lean.

The first manifestation of this difference in lean practice is that mechanical manufacturers can rely on materials that are generally stable throughout manufacturing and therefore materials that are tolerant of interruptions to the manufacturing activity, but process manufacturers cannot. Thus, for example, the lean practice of *andon*[5] or line stop practiced by frontline operators for quality improvement that is of great value to mechanical manufacturers *cannot be used at all* in most process manufacturing. In process manufacturing, when people stop chemical changes after they have been initiated, the product normally becomes worse—not better. The first distinction between process and mechanical manufacturing is *the nature of the changes that occur as a raw material becomes a finished product.*

Indirect Material Transformations

It follows that the second distinction between process manufacturing and mechanical manufacturing is *the manner in which those changes occur.* In mechanical manufacturing, the changes that occur in the raw material as it becomes a finished product are uniformly achieved quite directly by touching the raw material either personally, as in an assembly process, or mechanically, by using some sort of device such as a cutting tool or a wrapping machine.

By contrast, in process manufacturing, the raw material is normally touched only in order to contain it for processing or in order to create an environment within which the chemical transformation will occur. In many process industries, such as petroleum refining, the material of production is so thoroughly contained that, in normal operation, the process operators never see the raw material, the work in progress, or even the finished product. In process manufacturing, the material literally changes itself once the proper conditions for the reaction have been established.

This distinction in the methods of manufacturing leads us to understand that *mechanical manufacturers are relatively labor intensive* and *process manufacturers are relatively capital intensive.* As a result, process manufacturers must focus much more on the equipment side of lean than on the labor side of lean. This change in focus has both advantages and disadvantages in our lean practice.

Time as an Independent Element of Production

The final important difference between mechanical manufacturing and process manufacturing is that two elements of time, often described as "residence time" and "persistence," are necessary and independent elements of process manufacturing.

In mechanical manufacturing, the changes caused by touching the material are generally both instantaneous and permanent. For example, when the drill touches the engine block, the reconfiguration of that engine *instantaneously* begins. Further, any physical change that occurs in mechanical manufacturing is permanent. Consequently, it is easily possible to start and stop most mechanical manufacturing at any time or to vary the speed of manufacturing with few or no consequences. It is possible to drill half a hole today and finish it in the morning with no degradation in the quality of the hole or in the quality of the resulting engine. Equally, it is possible to assemble a lawn mower partially now and finish it next year and the resulting lawn mower will be no different in any way. Time is required to do the work in mechanical manufacturing, but time is not generally an independent factor in the success of the process.

In process manufacturing, the situation is quite different. Although some chemical reactions do occur at extremely high speed, all reactive transformations of material require a measurable amount of time—that is, "residence time"—for the chemical transformation to proceed from initiation to completion. Of equal importance, chemical reactions do not commence instantly and, once initiated, do not stop instantly. This characteristic is generally described as "persistence."

Residence time and persistence are important to us because many reactive processes only produce the intended product when the proper conditions for the transformation exist in a consistent and uninterrupted way for the entire period of the transformation. Starting, stopping, or otherwise altering the conditions of the reaction while it is in progress will not simply start or stop the reaction, they will noticeably alter the outcome and it is unlikely that the resulting product will be the one originally intended. When process operations are disrupted, something will happen, and it is almost certainly something that is not good. As a result of these time constraints, process manufacturers are often inherently less flexible than mechanical manufacturers. Therefore, they have both a greater need for and potentially can derive a greater benefit from the lean practices that enhance flexibility.

Case Study: A Simple Example in Process Manufacturing

Baking a cake is a visible and commonly observed example of process manufacturing and provides a good example of the impact of these differences.

- The high-density liquid raw materials of a cake are fundamentally different from the low-density solid finished product. There is no possibility that the baker can produce a cake by simply assembling the materials or by any direct form of labor. *This process is transformational rather than reconfigurational.*
- A cake results only when the raw materials are placed into an oven under the proper conditions for the appropriate period. Even the finest baker cannot produce a cake without an oven. *This is a capital-intensive activity where the manufacturing outcome is largely dependent upon the equipment.*
- When you first put the batter into the oven, it appears that nothing is happening. That observation is essentially correct. If you remove batter from the oven before the reaction temperature is reached, nothing permanent will have happened. There has not yet been even a partial transformation of batter into cake. *The batter is demonstrating persistence.*
- Now consider what happens if you only partially bake a cake or attempt to start a cake one day and finish it the next or in any other way disrupt the normal time of manufacturing. Substantially interrupting the transformation from batter into cake once the process has begun will always result in a permanently damaged and often completely useless final product. *The transformation requires uninterrupted residence time appropriate to the desired reaction.*

Each of these issues represents an important distinction between process manufacturing and mechanical manufacturing and each has a substantial impact on the way that lean is practiced in the chemical and process industries.

Special Case: Continuous Processing

Continuous chemical processes share all of the characteristics of the process industry generally and add the complicating factor of *continuity* because fresh raw material is constantly being introduced to the environment of transformation and material that has completed its transformation is constantly being withdrawn. The entire continuum of the transformation from raw material to finished product is continuously in progress within the same process.

Case Study: Continuous Processing

At Suncor, we extract bitumen from Canada's Athabasca oil sands and upgrade it into synthetic crude oil (SCO). The process for extracting the bitumen from the oil sand was first demonstrated more than 100 years ago. At that time, oil sand was placed in a pot of boiling water and, after some time had passed, the sand and oil separated. The oil floated to the top and the sand sank to the bottom. Although that simple batch-type process demonstrated that the separation was possible, it did not make it commercially viable. Remember that in order to be commercially viable the full combination of mining the sand, extracting the bitumen, and upgrading it to SCO—all under near-Arctic conditions—has to compete economically with the folks in the Middle East who simply put a hole in the ground and pump out oil.

Today's commercially viable oil sand plants each produce hundreds of thousands of barrels of SCO each day using continuous methods. In the continuous process mode, oil sand and hot water continuously enter the reaction vessel. Bitumen continuously floats to the top and flows over a weir to exit the separation cell. Clean sand and water continuously flow out the bottom. It is a great process and one that has enabled Canada to become one of the world's leading energy producers.

However, the continuous process is not without its unique difficulties. For example, if the sand and water slurry stops flowing or even if the flow rate slows below a critical velocity, the sand will fall out of suspension. At that point, instead of a pipeline of flowing slurry, we have a pipeline full of sand that will not flow again until it is taken apart and cleaned. Most continuous processes share this characteristic. They can enable a commercially viable business that would not otherwise be possible, but they are often surprisingly fragile when the continuity is disrupted. Therefore, as we consider the practice of lean manufacturing in the process industries, we will need also to consider the special circumstances of continuous process operations.

Lean is appropriate to process industries and can be applied very successfully. It is a great technology—one that I have deployed successfully for nearly three decades in discrete manufacturing and in process manufacturing. In combination with some other tools of improvement—a great strategy and an industrial culture that supports an engaged workforce—lean will allow you to achieve performance that your competition simply cannot match.

NOTES

1. American Productivity and Quality Center.
2. I use the term "small event" improvement to distinguish the actions of engaged individuals and teams at the frontline from the "big event" actions typical of engineers and managers. That distinction in no way implies that the improvements resulting from the actions of these workers are small. In fact, taken together, the business impact of many small event improvements is rarely small.
3. Donald Powell, senior vice president, Gilbarco (retired).
4. In my first book, *A Culture of Rapid Improvement: Creating and Sustaining an Engaged Workforce* (CRC Press, 2008), I discussed both business strategy and employee engagement in the rich detail that can be achieved only by devoting an entire book to the topic. Both of these concepts—strategy and engagement—are vitally important elements of business success as companions to lean practice. They are discussed in this book as part of the cultural enablers of lean, but there is much more information available in my first book that can be of great value to you as you adopt these capabilities in your business.
5. *Andon* is a Japanese term that refers to the warning signals on an assembly line when a defect occurs.

2

Lean Enterprise Thinking

INTRODUCTION

The Shingo Prize criteria specify two different categories of cultural elements that are essential to successful deployment of lean practices throughout your business. The prize board uses the first, ***consistent lean enterprise culture,*** to distinguish *lean theories and values* that are *well understood* and *utilized throughout* the business from lean that is simply being practiced as a few technical capabilities employed within the plant.

As part of this discussion, we will introduce the theory, concepts, and values that constitute *lean enterprise thinking* as well as the way that these concepts and values are best disseminated through *policy deployment,* which is the subject of Chapter 3.

The second Shingo Prize category of cultural elements is ***cultural enablers.*** We discuss those elements in Chapters 10 and 11. The cultural enablers of lean are *leadership and ethics* and *people development.*

The Shingo Prize board made quite a statement when it committed two of the four assessment categories to culture, particularly because lean is generally described by reference to the technologies that make it work. I believe that degree of commitment to culture is entirely appropriate.

Lean manufacturing was developed in the Japanese culture immediately following World War II, and in many important ways lean was especially appropriate for that culture. In fact, in the early days of Western implementation, a commonly heard concern was, "That might work in Japan, but not here." By the way, when I first began to deploy lean principles and practices in the chemical industry in 1987, I heard that phrase in a slightly different version: "That might work at General Motors but not at Exxon." As we will discuss, that concern is not correct in either form. But the culture of your business is indeed very important to the successful practice of lean methods because the technologies of lean are specifically intended to enable widespread participation in the improvement of your business, and that new participation implies many new and different social interactions.

At the time when lean was first practiced, Japan enjoyed a social culture where teamwork, respect for individuals, personal development, quality, and continuous improvement were powerful companions in all aspects of work and personal life. In that social environment, it was quite natural for a lean culture of highly engaged individuals to emerge at work in a way that enabled each individual to make his or her personal best contribution to the success of the team and of the enterprise. Further, Japan was (and still is) a very homogeneous culture, where there are few social issues with racial, religious, or other forms of the interpersonal differences that can disrupt team work in Western industry. Lean values were closely aligned with and enabled by the social values of the Japanese culture. There was no need to create a new culture for the workplace because the existing Japanese social culture fit these purposes quite well.

The positive attributes of the existing social culture that enabled lean to flourish in Japan are generally not present in Western society or in Western industry. They will not be present in our plants to facilitate lean implementation unless we specifically create a culture in our workplace that has those attributes. The emphasis of the Shingo Prize criteria on the cultural enablers of lean practice recognize that Western industrial leaders need to take proactive initiatives to create an industrial culture that is consistent with lean manufacturing.

Still, I have learned over the years that, although lean is especially appropriate to the Japanese culture, Western industry does have its own cultural advantages. When these are properly developed and incorporated into the culture of the business, we can produce an even more creative and productive result. One example of this is the cultural value prevalent in the West of gladly accepting the special contributions of naturally gifted people and incorporating them in their full richness to the benefit of the entire team.

Key idea: The Los Angeles Lakers basketball team relies on Kobe Bryant to score 50 points per game. The important cultural attribute of that special contribution is that his teammates do not resent him for achieving personal success or abdicate to him the responsibility of winning. They simply incorporate his special capabilities into their overall game plan.

Western industry manufacturing teams also routinely have natural leaders or naturally gifted people who can make special contributions in a way that would be very rare indeed in Japanese culture, where team values and consensus thinking are overwhelmingly mandatory. As with so many other aspects of lean, we have both a greater need and a greater opportunity to develop a lean culture that is appropriate to our task.

Thus, let us begin our discussion of lean values and the business culture that will allow you to realize the full benefit of your lean efforts. Creating your own version of this knowledge and disseminating it throughout your business is the essence of *lean enterprise thinking.*

DEVELOPING A WESTERN LEAN-ENABLING CULTURE

A lean-enabling culture is very different from the industrial cultures that are common in most Western plants today. Even among very advanced and successful businesses such as Exxon, many employees continue to believe that they work in a place that requires them to leave their brains at the gate. Most people have individual task assignments rather than assignments that yield team-based outcomes. Few people have any genuine autonomy to create improvement or the time to do so. As a result, few people in Western industry come to work expecting that they will make a personal contribution to the success of the enterprise beyond accomplishing the assigned tasks. Even fewer believe that they will do so as a member of an autonomous team.

Also, very real social changes are in progress that will transform the criteria of success for a team-based culture in Western industry. During my career at Exxon, I worked with the first black man hired as an operating technician, the first Hispanic engineer to become an executive, and the first woman to lead a refinery. All those changes occurred within the span of my career, and more changes are coming. In my last years with Exxon, the plant I led in Baytown formally recognized for the first time that we had employees with alternative life styles and that younger employees are seeking still more personal freedom as a balance to their work life. People are becoming more open about their religious and political preferences. All of these diverse attributes of Western social culture are finding their way into our plants in ways that can either diminish or enhance teamwork and our ability to create a strong industrial culture.

For these reasons, it has been necessary for the most advanced businesses formally and dramatically to make their work culture much more inclusive and to take a very proactive approach to valuing human diversity. Businesses that intend to benefit from widespread teamwork must intervene in the interpersonal working relations of their employees in substantially new ways in order to encourage people successfully throughout the business to participate fully in lean manufacturing or in any other form of improvement that requires small team actions, including widespread engagement of the workforce.

Full participation is truly necessary. As described in Chapter 1, at Exxon Baytown during the years when we were doubling the productivity of the business, we were routinely realizing more than 40 implemented improvements per person each year. That pace of autonomous improvement is more than 1,000 times the North American and European averages. No business with average employee engagement had a chance to compete with our pace of improvement.

Key idea: Throughout this book, I discuss the concept of "autonomous" work, which I define as work that frontline teams can conceive and implement on their own initiative and with their own resources. It is, however, work that is strictly bounded. The teams can autonomously do only work that corresponds to approved objectives and they can use only improvement methods for which they have specific authority. Even when doing the right things in the right way, they must follow rules of practice that enable effective but unobtrusive oversight by engineers and managers. You will find more detail on establishing and conducting autonomous improvement as we proceed. *By creating strict limits on what people have the authority to do autonomously, we simultaneously give them great freedom to do that for which they do have authority.*

Much of the benefit to be derived from lean practice comes from the contributions of people at the frontline of the business: people who are empowered by lean practices to make contributions that are not possible in traditional models of leadership and operation. If the culture of your business prohibits, restricts, or even fails to strongly encourage truly ubiquitous employee participation, then you will miss the greatest source of new improvement that lean could provide. *Lean enterprise thinking is a great enabler of autonomous improvement.*

Historical Perspective

At the heart of lean culture are the fundamental theories and concepts of lean. Because lean has been learned rather than developed in Western industry—and, more specifically, learned in Western discrete manufacturing—we in the Western process industries have less mature experience with the theories than do our counterparts in mechanical or Japanese industries. For this reason, it is useful to discuss the early development of lean in Western industry to illustrate current lean theory and practice as well as to describe the way that lean theory and practice have evolved.

Note: Although culture is important, I understand that for most business leaders the substance of lean practice is the most interesting part of the process, so that will constitute most of the content of this book. However, this introduction to lean values will provide a good understanding of lean theory as an essential foundation for the technical chapters describing lean practice.

Let us start this review at the time of the first introduction of lean practice to Western industry and look at the concepts of lean as well as the pitfalls that we experienced before we fully understood the application of these concepts. You may well find that you encounter some of these same pitfalls as you introduce lean in the liquids industry where it is still relatively unknown.

Key idea: A smart man learns from his mistakes. An even smarter man learns from the mistakes that other people have made.

Early View of Lean: Just-in-Time Manufacturing

The first Western use of lean manufacturing was likely in the automotive industry during the early 1970s. At the time, I was working for General

Motors and it was obvious to many of us that Toyota was doing something of great value that we could not match competitively. At that time, GM was in the same position as the competitors of Gilbarco and Exxon butyl polymers were (see Chapter 1). We knew the form of what Toyota was doing but we did not know the substance of that practice in a way that would allow us to reproduce its performance.

As we attempted to learn about the new manufacturing practices employed at Toyota, we made the classic mistake of observing only the form of their practice without understanding the substance that made it possible. As a result, we attempted to reproduce their lean practice by reducing inventories in our plants without understanding the operating changes required to make that form of manufacturing practice successful. As might be expected, when we removed the inventories, but did nothing of more substance, we experienced the full impact of the problems that the excess inventory previously accommodated. The poor results obtained as a result of this inappropriate initial approach delayed our ultimate adoption of lean by several years.

In those early days, Western industry generally knew lean manufacturing as "just in time." These words alone describe the fundamental problem that we had in understanding the concept. As it then appeared to us, the lean concept was straightforward. Inventory in all aspects was reduced so that raw material arrived from suppliers just in time to begin production. Work in progress at each location was manufactured just in time, as it was needed by the subsequent operation, and materials moved through manufacturing from one activity to another just in time to be used. Finally, finished product was produced just in time to meet customer demand. That concept sounded pretty elegant and self-explanatory, but we promptly learned that we did not know how to turn those simple concepts into reality. The very first thing we learned was that simply removing inventory often made our situation worse, rather than better.

As we gained lean experience, we began quite early to talk about "just-in-time" inventories as an intellectual contrast to "just-in-case" inventories. That is, as we began to experience lean, we began to understand that the problem was not inventories per se, but rather inventories held as an accommodation *just in case* a problem should arise. *The most fundamental of lean theories is that inventory exists primarily to accommodate the existence of other problems* and that, if the other problems did not exist, inventory would not be needed. The implied cause-and-effect relationship is critical to successful lean practice. To cause improvement, first remove the problem and then remove the accommodating inventory.

Consider, for example, the inventory that results from the problem of inflexibility inherent in manufacturing many different products on shared equipment. The time and cost of reconfiguring the equipment to transition from one product to another is generally well known, and it is fully understood that the losses due to reconfiguration must be recovered in subsequent production.

In terms of capacity, if it requires 4 hours to convert the equipment from one product to another and the market requires only 1 hour of production, then without producing excess material into inventory, the equipment has a productive utilization rate of only 20%. That is, 1 hour of production and 4 hours of conversion occur during 5 clock hours. However, if capacity constraints require that your equipment must be at least 80% productive in order to produce enough product to meet customer demand, then a 4-hour conversion time between products requires that each product must be in production after the conversion for at least 16 hours. In that case, during 20 clock hours, there will be 4 hours of equipment conversion and 16 hours of production.

The dilemma that arises from treating this as an inventory problem is obvious. If production is only sufficient enough to meet current market demand without inventory, then the plant is only 20% productive. If production is sufficient enough to make the plant 80% productive, then 15 hours of production must become inventory. The solution to this dilemma is to recognize that the problem is not inventory. The problem is that the plant is so inflexible that it requires a 4-hour transition time.

Still today, many managers and lean educators omit that logical step and believe that lean practice implies that inventory is inherently bad. In the mid-1970s, it took us years to learn the hard lesson that a focus on inventory per se is not correct. I strongly encourage you not to make that same mistake.

Despite the just-in-time imperative to reduce inventories, eliminating inventory alone does not correct the problem. The only solution to the problem of operational inflexibility is to reduce the time and cost of the transitions greatly. We now have the technology to do that as well as other technologies that resolve other forms of manufacturing problems that traditionally were accommodated by inventory. Each of those technologies will be described in its turn in Chapters 4 through 9. Together, the lean technologies allow lean theory to succeed.

Early View of Engaged Employees: Quality Circles

At the same time that we learned about the technical substance that enables lean concepts to be put into practice, we also learned that lean is essentially a tool for everyone. Managers and engineers do not have the time to develop lean practice in the detail required for its best and widest use, and that is not their best contribution to the success of the enterprise. *If lean is to be used in its full richness, then it must be used by people at all levels throughout the business—especially people on the frontline teams.*

However, in the early days of exploring lean practice in Western industry, just as we had difficulty comprehending the technical aspects of lean, we had similar difficulty when we tried to adopt the lean culture of problem-solving teams at the frontline of the business. Both of these difficulties arose from the same source: confusion over the difference between form and substance. In our first attempt to practice engagement at the frontline, we had only the form of lean culture without the substance that would make it effective. Only after we learned the hard way that the form we had adopted did not work without the substance did we later learn how to successfully engage our people to help us succeed.

Once again, the words we created to describe what we had seen in Japan misled us. We observed the Japanese practicing improvement at the frontline, and it seemed that periodically the Japanese workers stopped working, talked about their problems, and then—somehow—were able to solve them. We called these groups "quality circles." As it turned out, we were describing the form rather than the substance of quality circles.

At the same time and in the same way, the managers at General Motors who removed inventories and suffered as a result also began to get the operators together in our version of quality circles to talk about the problems. Again, we suffered from our lack of substance. We did not provide our frontline teams with any new capability to solve problems or to implement solutions if they found them. Thus, no problems were solved.

As practiced, quality circles in the mid-1970s were essentially face-to-face suggestion programs. Without real direction and with no new capabilities to make substantive changes, these sessions quickly became what one friend appropriately described as the "ain't it awful" meetings. People stopped working and rehearsed the problems of the day with no real ability or even intention to cause improvement. Because we had promised people that these sessions would make the work and the work product better, when we were not able to do anything of value, we disappointed many people in the effort and lost some more of their goodwill and commitment.

Key idea: The aggregate impact of engaging the multitude of people at the frontline is impressively powerful. But that power can easily get out of your control if you do not manage it carefully. A friend recently described the situation as similar to a person who believes that he can be a successful rodeo bull rider because he has seen it done and knows that he has to hold on to the rope. He does have to hold on to the rope, but if that is all that he knows and all that he does, when the bull is released, he is going to be very sorry indeed.

A real and substantive cultural change is required to enable effective direct participation of many people in the improvement process in precisely the same way that a real and substantive change is required in the technology of manufacturing to enable other lean practices. In order to engage effectively, people need clear goals and they need those goals translated so that they are meaningful to their own work.

In addition to their existing skills of operating the plant and equipment, people need the new skills of diagnosing problems and creating solutions. They need management to provide them with the time to practice those skills, with access to the resources that facilitate the improving changes and with a framework for action that will ensure the people at the frontline as well as management that all improvements are the result of doing the right things in the right way. The organization also needs a noninvasive form of technical oversight that assures everyone that the changes proposed are within permissible limits before they are adopted.

The substance of successful frontline problem solving takes far more than gathering people together to discuss problems or even granting people permission to make changes; management must affirmatively deliver meaningful capability for people to cause focused and well-controlled improvement. The attributes of engagement mentioned here are the objective criteria of teamwork that are required for engagement in industrial improvement.

In Western industry, we also need to create a business culture that draws people from many different social cultures into the same harmonious frontline teams that the Japanese achieve naturally. As we did with the technologies of lean, Western industrial leaders have now mastered the detailed knowledge of the cultural attributes of business that will allow us to achieve this result. (The formalities for creating and practicing a culture of frontline engagement are described in detail in Chapter 10.)

Eventually, of course, Western manufacturers did begin to understand both the form and substance of lean as well as both the form and substance of engaging teams at the frontline. Today, many examples of Western practice are at least the equal of their Japanese counterparts. Thus, let us now jump to the present and begin the discussion of current lean theory and practice as the basis for a *lean enterprise culture.*

THE EIGHT SOURCES OF WASTE

Lean practitioners have identified eight specific types of waste endemic to manufacturing and especially suitable for elimination through lean practices, including effective autonomous action at the frontline. As a result, lean implementation often focuses on "the eight big sources of operating waste." If you enjoy using the language of lean, the Japanese word for waste is *muda* and the eight sources of waste are often referred to as the eight *muda*.

1. *Overproduction:* Production of either finished product or work in progress that is not currently required by a customer or by the next sequential operation within your plant causes you to store productive capacity, materials, and labor that could otherwise be used for products that are sold immediately.
2. *Delay:* Delay occurs when the different parts of your operations are not connected and synchronized with the result that labor and capacity are consumed in waiting rather than producing.
3. *Transport:* Transport is material movement required by the manufacturing process that does not advance the material to a subsequent operation or move the product directly toward the customer. Transport adds cost without creating value.
4. *Process:* Manufacturing processes should be capable, reliable, and available. Processes should be visible, intuitive, and well controlled. Equipment and processes that do not meet this standard waste time, money, and opportunities.

5. *Inventory:* Inventory resulting from overproduction and inventory used to accommodate the existence of unresolved operating problems are a waste. Storing, managing, and owning inventory independently adds new costs. Inventory used to accommodate problems often hides those problems in a way that makes both the problems and the inventory that accommodates them permanent.
6. *Unnecessary work:* Operations need to be designed so that production is convenient and easy for the operators. Work or worker travel that does not add value to the product but is required by poor communication, poor access to tools or materials, or poor plant and equipment design is a waste.
7. *Defective products:* Production of defective products wastes capacity, labor, and material. Defective products that move through your plant cause waste to propagate into every operation they touch. Defective products delivered to the customer waste your reputation.
8. *Human capability:* A recent addition to the traditional list of seven *muda* is the recognition that without a culture of engaged people, much of the human capability of your business is wasted.

Key idea: The eight *muda* describe a tactical focus for lean practice that provides people with a shared understanding of where they can improve the business by driving out waste. The lean tools (described later) provide the mechanism for achieving that improvement. Many of these *muda* are mutually supporting. For example, as you improve the waste of overproduction, you naturally improve the waste of inventory.

The frontline tactical selection of appropriate *muda* for each team is directed by the overarching business strategy and the translation of the business strategy into local goals throughout your business. In practice, different teams will have different opportunities to contribute. The advantage of describing eight potential targets is that one or more of these opportunities for lean improvement should be available to each team regardless of the circumstances.

The eight *muda,* or the eight sources of waste, are an important part of lean enterprise thinking. The *muda* create a shared understanding of the sort of fundamental operating losses that are especially appropriate for identification and improvement at the frontline of the business. The *muda* describe *what* improvement opportunities people should be looking to capture.

THE RELATIONSHIP BETWEEN INVENTORY AND OPERATING PROBLEMS

One of the most fundamental values of lean manufacturing—indeed, the lean value that most people immediately think about—is the existence of a very real and very direct relationship between inventory, or some other form of excess resource, and operating problems. That is because inventory and other resources are typically used—knowingly or unknowingly—to prevent a manufacturing plant from experiencing the immediate impact of operating problems such as equipment failures, product quality defects, and inflexibility.

For example, work-in-progress inventory is frequently accumulated between sequential operations that suffer with unreliable equipment. By separating the operations with inventory in this way, either operation is able to carry on with production while the other is under repair. Absent such intermediate inventory, at any time that either unit shuts down, production would stop in both. Exactly the same relationship exists when using inventory to accommodate operations that suffer from product quality or other problems.

By using inventory to accommodate unreliable equipment or poor quality, the full extent and cost of the problem does not appear as an expensive real time crisis that stops production. Instead of experiencing unreliability as a real time crisis, the costs and impact of unreliability appear more gradually and more routinely as the cost of creating and managing the accommodating inventory. By distributing the otherwise instantaneous and intense cost of reliability or quality events across a long period of time and through the apparently uneventful activity of creating and handling inventory, the events themselves effectively disappear. In traditional thinking, the existence of this inventory enables greater productivity. In lean thinking, the inventory *accommodates*—in essence, masks—the waste caused by the original problems and adds the new waste of inventory.

Key idea: If you are able to resolve the fundamental problems with reliability or quality promptly, then immediate removal of inventory is possible. If you are not currently able to resolve the problems, removing the inventory without solving the problem is likely to make your situation worse (as illustrated in the discussion of the problem of inflexible equipment). Most importantly, *if you understand the relationship between problems and inventory, you can manage the situation to achieve the best possible outcome for your unique situation. That is the essence of lean thinking.*

For this reason, in lean theory, inventory is the *visible symptom* of manufacturing problems. Intentionally or unintentionally, inventory is normally created to accommodate the existence of unresolved problems. Inventory, therefore, has a very direct relationship to the problems that it accommodates. A lot of inventory in a particular place usually covers a large problem at that same place. A plant that typically has large inventories in many places probably has many problems throughout the business. *In the same way that the* muda *identify the nature of manufacturing problems, inventory identifies the location.*

Key idea: It is critically important for leaders who are introducing lean to understand this fundamental lean concept fully. Although inventory is itself a waste of resources, it is not generally the root cause of waste in manufacturing; inventory appears principally as a symptom of other problems. Recognition of the implicit sequence of cause (problem) and effect (inventory) is vital to success. Solve the original problem; then remove the inventory.

Despite the clear relationship between problems and inventory, the most common cause of the failure for new lean initiatives is that leaders and employees become confused about this fundamental issue. As you introduce lean to your operating teams, it is critical that both you and they understand that in order to achieve sustainable improvement, they cannot simply remove the inventory. Rather, they first need to solve the problems that required the inventory. As with many other operational issues, this is one of form versus substance. Removing inventory, *the form of lean,* is visible and easy. Solving basic operational problems, *the substance of lean,* is less visible and much more difficult. *Implementing the form without the substance does not result in sustainable improvement.*

Case Study: Lean Form without Lean Substance

In 2005, as part of a Suncor reorganization intended to increase functional alignment and expertise, management created an independent supply chain organization, which promptly initiated a campaign to reduce inventories on site. They called this activity "lean inventories." One of the inventories that they reduced was spare parts for equipment maintenance. Unfortunately, that inventory of spare parts existed because, in those days, Suncor had unreliable equipment and routinely needed those parts to repair it. As a result of this improper use of lean concepts, the inventory was removed while the equipment was still unreliable. Because we no longer had the parts to repair the equipment, Suncor's reliability problems became worse—not better—because of this initiative.

The eight *muda* describe the nature of the problems that lean seeks to address and inventory accumulation often serves as an indication of where those problems lie. However, as Suncor experienced with its supply chain organization, a misunderstanding of this concept, even one that occurs away from the plant floor and with good intentions, can have serious consequences.

VALUE STREAMS AND SUPPORT PROCESSES

Yet another part of "enterprise thinking" calls for businesses to demonstrate the application of lean values and improvement in areas beyond manufacturing. This turns out to be a surprisingly powerful and valuable requirement. Process industries are intensely focused on the operation and performance of our core equipment; thus, we generally have a pretty good idea of the opportunities for improvement that exist in those areas and, for comparison purposes, a good idea of how our competitors are performing in similar operations. However, because of this intense focus on the critical operations, we often lose sight of opportunities that exist in other parts of our business.

As we seek to use lean values and practices to engage everyone in improvement, we will find highly valuable opportunities for improvement outside manufacturing. These indirect opportunities have surprising potential for positive impact on costs and performance. Engineers who design and deliver new facilities can make them increasingly fit for that purpose; contract administrators can establish more supportive relationships with suppliers; supply chain managers can ensure adequate availability of spare parts; and support functions, such as Information Technology and Human Resources, can improve the morale of the workforce through improved performance. All these and more provide real improvement opportunities to the entire business.

Case Study: Nonmanufacturing Improvement Potential

At Exxon in 1987, I encountered a classic example of just such an opportunity. At that time, when an employee relocated from one city to another, a small group inside the human resources organization was charged with administering a self-insurance program for losses or damage to the employee's personal property that occurred during the move. At that time, Exxon was spending in excess of $70,000 each for domestic moves to relocate executives very gracefully, with the goal that the move would be so uneventful that the executive would become immediately productive in the new location. That money was well spent. Executives routinely arrived at their new location and fully engaged within a few days following the announcement of the transfer—that is, unless there was important damage to their personal property during the move, at which time the HR self-insurance group became involved.

Their best elapsed time to process a claim for damages exceeded 6 months and some claims required far longer. In comparison, the Texas Insurance Code specified for the regulated insurance industry that a delay of 90 days constituted "prima facie evidence of intentional misconduct." To be specific, in this situation the best efforts of the insurance administrators in the Exxon HR organization were twice as bad as performance that the legislature and industry professionals considered to represent intentional misconduct. These were not bad people. They were simply people who were doing work that was so far from the mind of management that literally no one recognized the extent of the poor performance. However, the outcome was important because people upon whom we counted to hit the ground running in a new location were frequently distracted for months by unresolved personal problems.

More recently, I encountered another memorable example of the opportunity for improvement away from the plant floor. Suncor became very cost conscious when the price of oil dropped from more than $150 per barrel in the summer of 2008 to less than $40 per barrel in January 2009. Among other things, we sought to reduce a relatively large expenditure on company-owned light vehicles (cars and pickup trucks usable on the public streets).

Despite several weeks of serious talk with senior managers, it appeared that they all considered every vehicle critical to the operation. One day, the supervisor of our snow removal team appeared at my office door with a list of 147 vehicles that had not moved during both of the two most recent snowfalls. With that information in hand, we were able to break the back of the resistance and we ultimately saved more than $20 million in annual costs for light vehicles. It was a great reminder to me that *when everyone knows precisely what you are trying to do, everyone can help.*

Key idea: Throughout any organization, there are always a great many examples of this sort that can provide savings or improved performance or simply demonstrate to everyone that *all parts of the company are on the same path toward excellence.*

The lean tools apply to improvements in the indirect areas of the business in precisely the same way that they apply to the work in the plant. The eight *muda* exist in the front and back offices just as they do on the plant floor. The critical issue is to organize the people who work in the indirect areas to think about their work in the same disciplined manner that we think about work in the plant. A manufacturing tool that is especially useful in this regard is "value stream mapping."

In either manufacturing or indirect operations, the essential concept of value stream mapping is to identify clearly the value to be created and to understand clearly each element of work as a contribution to that value. In manufacturing, this practice is used to identify work that does not contribute to the intended value of products and therefore is an appropriate target for waste elimination. Value stream mapping has precisely the same effect for indirect work. Further, in indirect areas, value stream mapping may be the first time that the people performing that work ever stop to consider formally the *intended value* of what they create as differentiated from the routine tasks of their assignments. The assessment and the knowledge gained in that assessment often have surprising impact.

Case Study: The Value of Value Stream Mapping

I once helped Exxon Chemical's vice president of technology with a problem in one of Exxon's technology centers. Although this was purely a research activity in a facility dedicated purely to science, it proved Thomas Edison correct when he observed that invention is "10% inspiration and 90% perspiration."

The problem involved PhD-level scientists but it soon became clear that the nature of the problem was activities that looked more like perspiration than inspiration. The development of new molecules (business value) was being harmed by waste in the planning, scheduling, and execution of standard laboratory support activities (value stream). Because the scientists believed that their only important function was to be creative, the other work, which enabled their creativity, was not well organized or managed. When we understood this issue and got the routine work right, the creative process proceeded much better. Value stream mapping, or understanding the flow of value as it relates to the flow of work, can help you with efforts of this sort.

There are many great stories of surprising improvements in the indirect areas of business attributable to lean thinking and value stream mapping. As you learn about applying lean tools to the manufacturing functions, you should remember that they apply equally in other areas as well.[1]

LEAN VALUES: INVENTORY REDUCTIONS CAN SUSTAIN IMPROVEMENTS

Three additional lean values associated with the relationship between resources and waste will help you further establish a successful basis for lean enterprise thinking in your organization:

1. *When you have successfully resolved problems, you need to be certain to remove the inventory that previously accommodated those problems* to ensure that the solutions you have created are *permanent.* If a problem that has been resolved returns (as problems often do) when there is no longer an accommodating resource, reemergence of the problem will be immediately and intensely apparent and you can promptly remove it again, paying greater attention to creating a permanent solution.

   Similarly, the recurrence of an accumulation of inventory after you have solved the underlying problem and removed the original inventory also indicates that the solution you have put in place may not be permanent without further effort. Removing the resources that previously accommodated a problem that has been resolved is not only an additional source of improvement but also a great way to sustain the solution.

2. Knowledge of the cause-and-effect relationship between problems and the accumulation of resources is valuable when you are not sure where or what your problems are. *Identifying and examining the accumulations of resources in your plant can provide great evidence of an underlying problem that might not otherwise be easily recognized.*

   This is a primary benefit of enterprise thinking. I have experienced some surprisingly valuable outcomes when people throughout an organization began to question small but unnatural accumulations in their areas. (This will be covered in some detail in Chapter 6.) In large organizations, many small but costly problems are accommodated by small but valuable resources. Those problems might not attract the attention of engineers or managers, but they can be identified and resolved by people who are close to the issues.
3. Inventory often can be strategically deployed in a way that enables you to accommodate problems that you cannot immediately solve. When you practice lean manufacturing, *your goal is to fix problems, rather than to experience them.*

Case Study: When Inventory Accumulation Is the Only Choice

In 1986, I toured the Nakayama Steel Works plant near Kyoto. Until we went outside, it seemed to me to be one of the best examples of lean that I had ever witnessed. However, once outside, I saw an enormous pile of coal. At that time, Nakayama kept a 4-month supply of coal on site. Even to my very Western eyes, that was not a lean inventory of raw material. The underlying problem was that, between Nakayama and its supply of coal, were the Pacific Ocean and several railroads as well as mines operated by the then notoriously erratic Australian trade unions. This "best of the best" lean operator had recognized a problem that it could not fix or control, so it buried it under a pile of coal. Because the company could not fix the problem of unreliable coal supply, it made certain that it would not experience a shortage, which would disrupt otherwise excellent operations.

LEAN VALUES: CULTURE OF ENGAGEMENT

An essential lean value is that everyone in the enterprise should be engaged personally and directly with the success of the business. The lean theory enables people to recognize waste in your operation. The lean technologies enable your people to remove the problems and the waste that accumulates around the problems. In the process, they learn more about how the plant works and how they can operate it well and make it better as they do. Work becomes a single integrated activity for operating and improving the business.

You should anticipate that, as you employ lean theories and lean tools in a culture in which everyone helps by contributing their personal best, you will discover productive capacity far exceeding what you now believe you possess along with the capability to do new things that you do not now imagine to be possible. Your products, delivery, and service should all become better and, of course, your costs, productivity, and profits should all improve dramatically.

The benefits of creating an engaged workforce have long been recognized, but I have met a great many leaders who have failed in the attempt with sufficient regularity that they have now essentially abandoned hope of achieving that outcome. The problem is that these leaders imagine that their employees are bristling with important contributions of immense value and are just waiting for the leaders to "get out of the way." Nothing could be further from the truth and nothing could be more detrimental to the people and the business than attempting to practice employee engagement by simply granting people permission to implement unstructured change.

People do have a great potential to contribute in a more fundamental way and they do have a great desire to do so. Lean theories, tools, and practices will give them the practical capability to do that. However, enabling many people to make individual contributions without the chaos that would likely result from many random actions requires a disciplined management effort. As a business leader in a process industry, you do not want and indeed cannot tolerate random changes, including changes that are intended to be improvements. Liquid manufacturing is a serious business that often is severely or dangerously disrupted by unstructured change.

We want engaged people who continuously do the right things in the right way, using the right tools and staying within the boundaries that define safe and appropriate operations. In short, we need people to play their role as part of a disciplined and focused industrial team. The greatest value contributed to a team effort by all people, including natural leaders or naturally gifted people, is playing the role defined for them as part of the team. Effective teamwork begins with defining the mission of the team and the roles of individuals within the team and communicating the rules of the game. Creating the shared understanding of *lean enterprise thinking* is an important cultural element of this work; however, the development of an appropriate industrial culture also includes the strategic planning and policy deployment described in the next chapter and the other elements of culture described in Chapters 10 and 11.

Key idea: To obtain full value from your lean efforts, you need to establish an industrial culture in which *everyone* in the enterprise is highly engaged in contributing to the success of the business. Lean enterprise thinking and the other cultural enablers of lean represent a foundational capability for lean practice. Far different from the concept of "getting out of the way," creating and sustaining an engaged workforce is the most disciplined and difficult—but also most rewarding—management assignment you will ever undertake.

NOTES

1. The lean practice of value stream mapping in the process industries is essentially the same as in the mechanical industries, so it will not be discussed further in this book. For this reason, I will not spend more time discussing other opportunities for improvement in the indirect areas of the business. Human resources and other indirect functions in the process industries look just like HR and other functions in the mechanical industries. In this book, we will focus our attentions on those areas where process manufacturing needs special treatment or where I can provide you with specific examples from our industry.

3

Policy Deployment

INTRODUCTION

The second element of a ***consistent lean enterprise culture*** is *policy deployment.* At the most fundamental level, lean enterprise thinking provides your team with an understanding of lean theory and values, which allows you to develop your own approach to making those values a reality in your business. Policy deployment, as a natural companion to shared values, creates a sharp focus for ubiquitous shared action. Together these two elements of lean enterprise culture enable you to engage many people at the frontline, who can use lean theories and lean tools to make contributions that would not be possible in traditional operations.

Lean is an inherently powerful contributor to your manufacturing enterprise and you can successfully benefit from those capabilities through management-led initiatives as well as through the efforts of your engineering organization; however, you should remember that lean is most beneficial when you enable frontline employees to contribute. The full capability that lean offers to business leaders cannot be achieved unless you take the steps necessary to engage and enable many people to help you *recognize* opportunities and, of greatest importance, provide autonomous implementation to help you *benefit* from those opportunities.

In successful lean businesses, more than half of the total improvement realized comes from people who never previously contributed to improvement in any manner. Shared understanding of both the objectives of the business and the practices of lean enables people at the frontline to work in complete harmony with the existing capability for improvement represented by your engineers and managers.

Although for decades business leaders have sought improvement from the teams at the frontline of the business in many different ways, that aspiration has generally been unfulfilled. The principal reason that this highly desirable goal has been so difficult to attain is that we normally limit the scope of engagement available to people at the frontline to identifying problems and nominating them for others to solve. In that way, the total corporate capacity for improvement remains limited by the technical and execution capacity of your existing professional improvement staff. In fact, because people at the frontline are generally interested in problems that do not attract or merit the attention of engineers and managers, responding to the problems nominated from the frontline often distracts from larger work that the traditional improvement team could otherwise be doing.

As an alternative to suggestion programs, managers have periodically sought to obtain autonomous improvement at the frontline by simple request or fiat but have failed to provide a meaningful structure to ensure that the actions at the frontline are both strategically correct and within the bounds of business necessity. The results of such unsupported efforts are always modest because many people at the frontline do not know what actions will produce strategically correct improvement or how to achieve industrial improvement effectively and therefore fail to engage in unstructured autonomy. Because improvement is only modest and there is a strong possibility of misdirection, this form of engagement is not worth the risk to the business or to the individuals and normally is promptly abandoned when the first autonomous effort goes substantially out of bounds.

Key idea: Policy deployment is the management approach to aligning the many independent activities of autonomous teams so that they are compatible with one another in a way that, together, they safely produce significant aggregate progress toward the focused objectives of the business.

LARGE EVENTS AND SMALL EVENTS

If you truly intend to get full benefit from a new capability for improvement at the frontline, you need to recognize formally that, in manufacturing, two distinct types of improvement opportunities are affected by two very different types of solutions. Engineers and managers conduct "large event" improvement that demands corporate-scale resources and requires a relatively long time to implement. People at the frontline conduct "small event" improvement that normally can be accomplished in a short period using only the more limited resources that are naturally available to them. By describing these events as large and small, I am not characterizing the impact of the improvements, but rather the scale of the resources and people engaged in the effort and the time required to create the improvement. Small events often have great value, especially when aggregated.

The value of mentally segregating improvement into large events and small events is that, when the goals for small events are carefully focused and the boundaries authorized for small events are appropriately managed, people at the frontline can implement small event improvements on their own initiative (autonomously). It is important to note that, as a result of the beneficial structure provided by the focused goals and careful limits of small event improvement, this approach is, as a rule, much more valuable than unstructured autonomy. People can independently contribute *completed* improvements that are strategically aligned and certainly within the needs of the business. Frontline teams will not make the same improvements in the same way as engineers do, but they can make their own improvements in their own way as an important companion to the efforts of the engineers.

In fact, when policy deployment is well done, the small improvements at the frontline are often so complementary to the efforts of the engineers that the engineers are able to provide more frequent and better big event improvements. As we discuss policy deployment, we will do so with an eye toward engaging people at the frontline in small event improvement work as an integrated part of the entire improvement team.

A STRATEGIC VIEW OF MANUFACTURING

Many business leaders become so consumed by carrying out the tactical details of their routine activities that they fail to create a strategic understanding of the business and communicate it to the people who help them run it. Similarly, many manufacturers become so lost in the details of routine operation that they fail to communicate a strategic understanding of the manufacturing process to their people. As a result, it is virtually impossible for most people, especially those at the frontline of the business, to have visibility of objectives or opportunities beyond completing the immediate task.

When lack of vision and understanding limits people to rote conduct of a given task, they have very little ability to recover from disruptions to the routine and essentially no ability to make innovative and meaningful enhancements to the work. Lean enterprise thinking provides people with a strategic view of your manufacturing process and policy deployment provides them with a strategic understanding of your business and improvement goals. When people understand how you intend to work and what you intend to achieve, they have much greater ability to make innovative contributions.

Key idea: This situation at the frontline of manufacturing is generally similar to that of drivers who receive "turn-by-turn" navigation instructions, but no map. If the route to their destination is exactly as expected, then that basic information is often sufficient. However, if there is a detour, accident, or just a traffic jam, the folks who have nothing but turn-by-turn instructions are unable to recover from the disruption: They are either hopelessly lost or stopped. Further, even if the intended course is clear, when people have no understanding of the intended destination and no map for reference, they have no ability to find a better alternate route that was previously unobserved.

In addition to limiting the information available to people, managers often impose unstated constraints on activities at the frontline because we assume that people necessarily will do things as we would do them (large events). *Very often, people can do things in a much simpler and better way (small events) if they know the destination and it is clear to them how to get there.*

Case Study: A Business Trip to Fort Calgary

I recently needed to travel from Suncor headquarters to Fort Calgary, a nearby historical museum that also provides rooms for events and conferences. A helpful administrative assistant printed off a copy of the turn-by-turn directions for me using a popular online mapping service. Fortunately, Calgary is a very logical city with downtown streets largely organized into quadrants and designated sequentially. As a result, it was very easy for me to understand quickly how to get from Suncor on 4th Street to Fort Calgary on 9th Street.

Having a clear destination and an easily identifiable path was indeed fortunate because I was walking and the turn-by-turn directions assumed (without saying so) that I had a car. As a result of that unstated but embedded

FIGURE 3.1
Servicing equipment from a scaffold.

assumption, the turn-by-turn directions intended for me to follow a route that considered the effect of one-way streets as well as intersections that prohibited left turns. That information was not appropriate for my trip. On foot, I was able to ignore those constraints. There are no one-way streets for pedestrians and we can cross at any corner. I arrived at my destination faster than if I had been in a car. Importantly, I arrived *very much* faster than would have been possible if I had attempted on foot to follow directions that were intended for drivers.

Figures 3.1 and 3.2 illustrate one of my favorite examples of operators who have found their way around obstacles that management creates. Suncor management has created very specific standards for servicing equipment, such as the pipe rack, at elevation, using a scaffold that is shown in Figure 3.1. That management-imposed standard has been in place for a very long time.

Case Study: Circumventing Obstacles

Our frontline teams recently found another way. Instead of building a scaffold platform from the ground up, they used mountain climbing gear and descended to the job from above (Figure 3.2). The total cost reduction obtained by substituting ropes for scaffolding exceeds 80% each time this technology is used. Further, the delay in starting the work normally caused

FIGURE 3.2
Servicing equipment from above.

by constructing the scaffold is effectively eliminated. This technique is not appropriate to every task, but when it is appropriate, it is perfect.

STRATEGIC ALIGNMENT AND NECESSARY BOUNDARIES

In the chemical and process industries, allowing people at the frontline to make autonomous improvements (improvements imply change) requires that we construct and operate a very disciplined framework. This is done so that we can be certain that everyone is consistently doing the right thing, in the right way, and well within boundaries that define clear limits on the nature of the changes that can be made without engaging the normal processes for management of change. For this reason, the discussion of policy deployment in the process industries will be very rigorous when compared to a similar discussion applied to mechanical manufacturing.

This is another situation where the opportunity for benefit in process manufacturing is much greater than it is for most mechanical manufacturers. Because the consequences of unmanaged change in our industry can be so severe, we have not traditionally allowed much change of any sort that is not the work product of a highly disciplined engineering effort. In fact, in the process industries, we have discouraged people at the frontline from autonomously making any changes of any sort. As a result, Western mechanical manufacturers typically do not receive much improvement from the frontline when compared with the best manufacturers, and process manufacturers typically receive far less than that because we have such a concern about change management.

Whereas mechanical manufacturers can often simply turn people loose to make improvements, we need to construct and manage a rigorous process for ensuring that engineering controls are uniformly maintained. For autonomous improvement, this control is typically achieved by establishing and monitoring clear boundaries within which we can be certain that people can work safely. With such a rigorous process, we can receive abundant frontline improvement equal to that achieved by the best businesses. Carefully defining the limits within which it is certain that people can do the right things in the right ways means that they can achieve great improvement within those limits.

The social model used by large groups to ensure that people routinely do the right things in the right ways and within well-understood limits is known as culture. Policy deployment and lean enterprise thinking form the basis for an industrial culture where employees all share the values, beliefs, behavior, and rituals of the enterprise. These attributes of industrial culture are identical to the fundamental attributes of a social culture and they define mutually reinforced standards of conduct for how people act within that culture.

Values (strategic goals) and beliefs (tactical goals) in an industrial culture describe the theoretical and practical alignment of everyone toward the focused goals of the enterprise. *Behavior* describes the specific actions that are expected and authorized and the limits on those actions that will be enforced both formally and socially. Finally, the *rituals of an industrial culture,* such as the quality stations described later in this chapter, unite the full team at work in the same way that rituals unite members of other cultures. "This is how we do it" is a unifying statement that has great value to your enterprise. When you engage everyone in a shared culture of improvement, everyone can help you ensure that the outcome of that culture is as you expect it to be.

Key idea: We all know how to form small teams of five or ten people for games or other social purposes. In most situations, we develop ad hoc cultural attributes—values, beliefs, behavior, and rituals—for the team in addition to the existing attributes of a social culture that are likely already shared by the team members. For industrial purposes, we typically have much larger groups, often including people from many different social cultures. When we create an industrial culture based on values, beliefs, behavior, and rituals that are unique to the needs of the business and uniquely appropriate to the people working in it, we can transform such a large and diverse population into a unified team at work.

PREREQUISITES FOR STRUCTURED AUTONOMOUS IMPROVEMENT

Policy deployment consists of three separate elements, each of which is equally important to the creation of autonomous improvement. The first element is *strategic direction.* If we want people to have the autonomous ability to help us with meaningful improvements that all add to and are compatible with the improvements provided by others, then we need everyone collectively to pursue a few focused objectives that will have the most value to the enterprise. In the process industries, it is equally important that everyone share the same understanding of what changes are *not* authorized for autonomous action. Leadership must be able to articulate the business values and the strategic intent *clearly and consistently* throughout the enterprise so that everyone shares the same goals and the same boundaries. *Shared strategic intent is a basic element of creating an engaged workforce.* Authorizing people to make changes without a clear understanding with which to organize, direct, and limit autonomous effort invites random actions that are often wasted or, even worse, chaotic.

The second element of policy deployment is *translating the goals and values* of the business to make them meaningful at each place where people will work to transform the goals into reality. The words and concepts typically used to describe the strategic goals of the enterprise in a way that is meaningful in the boardroom are often unintelligible at the frontline without this translation. For example, the CEO will immediately understand a corporate goal to increase the capacity of the plant without new investment. However, that goal provides relatively little direction to a mechanic working alone at midnight on Sunday in a bitumen extraction plant. A derivative—or translated—goal appropriate to the extraction business might be to increase the productive capacity of separation cells by reducing the time consumed by outages for routine maintenance. That derivative goal accompanied by appropriate tools of improvement tells the mechanic precisely how to help.

In addition, as the goals of the business are translated throughout the enterprise, it is important to establish both vertical and horizontal alignment among the several frontline teams. There are many occasions in a complex organization when the actions of one team affect others. Teams need to know what is important and what is prohibited. They need to understand the impact of changes in their work on the work of others. *An important boundary for autonomous action is that teams cannot be permitted to make their own work better in ways that make the work of others worse.* Compared to a traditional structure of change led by management, change led by people within individual teams takes on a completely new intensity when it is related to frontline visibility of and respect for the work of other teams.

The third element of policy deployment is the establishment of a *framework for action* that will make the improvement process, as practiced by each frontline team, visibly apparent. In that way, people can continuously demonstrate that they are making progress and that they are doing the right things in the right ways while staying within the limits. In addition, management can be certain that, even with many people taking many actions, the results are all positive and the actions of all teams are mutually additive and compatible, well controlled, and consistent with ordinary practices for the management of change. When the actions and intentions of all frontline teams are visible, members from each team have the opportunity to learn from other teams and management has a noninvasive means to monitor the work to ensure that the objectives and boundaries for autonomous action are carefully observed. Visibility also makes it apparent that each team is indeed actively practicing frontline improvement.

An interesting attribute of an industrial culture of improvement based on commonly held values, beliefs, behavior, and rituals of the enterprise is that important social rewards (or consequences) are derived from doing (or not doing) the right things in the right ways. This social effect can currently be seen in many of the best chemical operations that practice workplace safety observations. By observing each other as we work and either complementing or correcting safety behavior, many people *immediately improve* our safety performance and we *continuously reinforce* our culture of safety.

We will discuss each element of policy deployment separately and then examine what the completed system looks like once implemented.

STRATEGIC DIRECTION

In the lean context, the key element of strategy is not deciding whether your business will enter new markets or develop new products, the key element is to give people who would not otherwise be engaged with changing or improving your business sufficient information about your strategic direction to allow them to help you succeed. Strategy here is largely about the clear communications that are critical to successfully *sharing the work to achieve the strategic intent.* When you communicate your strategy, you are providing people with a clear view of your destination. Compared to the current state of most businesses, simply letting people know where you are going represents a massive improvement over "turn left at the corner."

Lean is not participative management where everyone decides what is important. Lean is participative execution where everyone contributes to achieve the established goals of the business. The first step in that process is to ensure that everyone knows what the goals are and how he or she can help.

With the intent of ensuring that people newly authorized to make autonomous changes are consistently doing the right things in the right ways and staying within the boundaries, your strategic direction defines the *right things* to do. A good strategy for these purposes makes as much impact by defining and excluding what is *not* authorized as it does by defining and including what *is* authorized. In this way, strategies provide both direction (*toward* the right things) and focus (*away from* other things).

Let us assume that you adopt a strategy intending to *increase the capacity* and *reduce the costs* of *existing* products and operations with *no new investment.* It is clear how that strategy can be of long-term benefit to the business, and you can imagine how to translate that strategic intent into meaningful actions for frontline teams throughout your plant. People can receive and understand such a strategy in a way that will enable them to implement a great deal of improvement in many ways and in many places, all of which will add to and be compatible with the actions of others.

With such a strategic direction, people can, with confidence, autonomously make operating improvements that have *no other effect* than to lower cost or increase capacity. There are plenty of those opportunities.

Providing clear strategic direction and focus does not reduce the number or the scale of the actions available to your team. Direction and focus simply get the entire team moving together in a way that ensures all actions are aligned and compatible.

At the same time, this strategy is clearly limiting: By specifically restricting new investment, it requires creativity before capital. It is specifically clear that new products and new facilities are not within the strategic intent. People have no authority to make product quality worse in order to increase capacity. They have no authority to make product quality better in a way that reduces capacity or increases costs. They have no authority to launch a capital project even if it will increase capacity or reduce costs.

The important issue for management and for the teams is that the specific types of improvements that are authorized for autonomous action are very clearly defined. Improvements not specifically authorized for autonomous action are excluded explicitly or implicitly. When both the directions and the limitations of the strategic intent are clear, people at the frontline as well as managers and engineers know that those pursuing the shared goals are doing the right things. In this way, you establish the values of your new industrial culture. Note that in addition to the limitations on the "what" of improvement that are provided by the strategic intent, there will also be limitations on the "how" of improvement.

Key idea: An important limit on the autonomous activities of frontline teams in the process industries is your formal practices for "management of change" and the other elements of process safety management. Autonomous improvement teams must always follow the same rules as everyone else. Autonomous action is not a license that allows frontline teams to avoid the strictures that ensure process safety. Different people throughout your business will take different actions in different ways, but everyone must follow the rules when making changes or improvements. Management of change and other rules of practice form the boundaries within which people at the frontline work as they practice improvement.

At both Exxon and Suncor, we disallowed any *autonomous authority* for actions such as penetrating a pressure boundary or a hydrocarbon containment envelope that would otherwise fall within our formal management of change procedures. Only engineers or others with *specific formal authority* could make those changes. Despite that significant exclusion, by clearly establishing what was permissible, we obtained a tremendous amount of valuable improvement. *By carefully defining what people could not do, we and they were easily able to recognize what they could do.*

The strategic boundaries for autonomous frontline improvement are not absolute limitations on the actions of the corporation. This strategic direction only limits the changes that those at the frontline are authorized to make autonomously. The leaders of your business and even the frontline teams may elect to undertake other initiatives in other ways by complying with the formalities of change management much as they do now.

Therefore, your first step in policy deployment is to communicate what needs to be done, what should not be done, and the boundaries that cannot be exceeded. In the context of lean manufacturing, this very pragmatic process focuses on communicating clearly with the people you want to engage. You will find that your people can make a great many contributions if you do a good job of describing what you want.

Case Study: Communicating Goals

As we undertook this process at Suncor, we focused our frontline teams very carefully on improvements associated with improving the "safety" and "reliability" of our operations. At the management level, we had companion goals associated with "people" and "environment," but for most purposes, as we translated our goals to the frontline teams, we focused principally on improvements in reliability and safety. As policy deployment progressed, we discovered that the frontline teams had far more capability for making a meaningful contribution to improving our environmental performance than we had originally thought, and we adjusted our strategies at the frontline in order to include them in that effort. You are also likely to find that your first attempt to translate the strategies of your business broadly is imperfect. That is fine. Lean is an ongoing process of improvement for all people, and managers are people too. It is perfectly satisfactory to improve the goals—especially the translation of the goals—as you proceed.

THE ROLE OF COMMUNICATION IN ACHIEVING STRATEGIC ALIGNMENT

The concepts and practices for deciding upon the strategic goals of your business are well documented elsewhere, but the communication of strategic goals throughout an enterprise as the foundation for a ubiquitous practice of autonomous improvement is not as well documented; therefore, that will be our focus here. In most businesses, it is common for people at the front-line to receive little meaningful communication regarding the strategic direction that has been set by management. Communicating the essence of your business strategy can significantly help you create alignment as you implement lean and seek more improvement from more people.

Key idea: Even the very best businesses lack the ability to improve everything rapidly. Great performance comes from *strategic focus,* which *allows all improvements to contribute to achieve significant progress toward a few very important objectives.*

The first critical communication element in policy deployment is to be certain that you are in fact deploying strategic goals. By a very wide margin, the most common mistake that leaders make in this regard is holding the strategic intent of the business at the management level, while deploying to the frontline only specific activities that they hope will produce the desired strategic outcome. If people do not know what you are trying to achieve, their ability to help is substantially limited.

Key idea: As you use the goals of your business to communicate how your employees can help the company succeed, every frontline team needs to share your strategic intent. *Strategies always describe business outcomes rather than activities intended to contribute to those outcomes.*

Ensuring continuity of strategic intent is not as complex as it may sound. Let us look at a simple chain of strategic intent:

- At Suncor, one strategic goal for the corporation is to increase our productive capacity for synthetic crude oil (SCO).
- Because bitumen is the raw material of SCO, that corporate goal translates directly to a strategic goal to increase our capacity for bitumen production.

- In our primary extraction plant, that further corresponds to a strategic goal to increase the capacity for chemically separating bitumen from sand.
- In our secondary extraction plant, that corresponds to a strategic goal of increasing the capacity for separating mechanically entrained sand that remains after the chemical separation has been completed.
- For the maintenance team that serves the cyclonic separators, that goal corresponds to a strategic goal to improve the mechanical availability of the units.
- For the preventive maintenance team that serves the cyclones, that goal leads to a strategic goal to improve the success of preventive maintenance while reducing the duration of operating outages for routine maintenance.

As this simple translation demonstrates, even at the frontline of the maintenance organization it is possible to define goals that have strategic intent directly linked to the strategic intent of the business. It is apparent to the frontline teams that maintain the cyclonic separators that this strategic goal for the Suncor business also has attributes that are strategically meaningful to their work. The maintenance team does not simply have routine tasks to perform; rather, they have strategically important business outcomes to achieve. As the teams undertake autonomous improvement, the impact of knowing the strategic intent of their work is invaluable. With strategic objectives in mind, teams can properly undertake many tactical goals and actions that will contribute to the success of the business. In this way, many people can contribute many things over an extended period in direct support of the goal.

However, many teams today only receive tactical goals and that is much less valuable. A *tactical goal,* like a strategic goal, generally relates to a large body of work that allows many people to do many things over an extended period. The critical distinction is that tactical goals only have value to the business when they support strategic goals. When the strategic goals are unknown, it is easily possible for tactical goals to deviate from the original intent. For example, in the extraction plant, maintenance improvements that contribute directly to bitumen output certainly have strategic business value. But tactical maintenance improvements that contribute principally to the professional conduct of maintenance—for example, those that increase compliance with the maintenance plan—may have no business value at all.

That is why it is necessary to communicate the strategic intent "to improve maintenance in a way that increases bitumen output" and not the tactical intent of "improving maintenance" or the more focused tactical action of "improving compliance with the maintenance plan." General improvements in maintenance practice, such as improving compliance with the maintenance plan, may have value, but they do not have independent strategic value. Those tactical goals produce business value only by contributing to the strategic intent and that value is only ensured when a direct linkage to the strategic intent is maintained.

Business leaders who make the common mistake of deploying only tactical goals to their frontline teams normally do so by making the link between strategies and tactics at the managerial level. The leaders who deploy tactical goals are thinking strategically, but instead of deploying strategic goals, they deploy only the tactic that they believe will produce the intended strategic result. Tactical goals and actions are an important part of implementing any strategy, but *you must always be willing to alter the tactics if you find that they are not sufficient to produce the strategic outcome or if you find that they are not properly aligned with the strategic intent.* When the tactical goals have been directed by management, frontline teams who do not know the strategic intent cannot recognize the need to make that change and they likely do not have the authority to do so. Meanwhile, managers may not notice the problem in a timely manner. As a result, it is surprisingly common in this situation for teams at the frontline to pursue clearly inappropriate tactics for long periods.

Case Study: Strategies versus Tactics

In the cyclonic separation process, Suncor management had recognized that increased operational availability of the equipment was strategically important. Unfortunately, they adopted a maintenance approach that they believed would produce the strategic goal of improved reliability, but that approach consisted of deploying only the tactical objective—*improve planned maintenance in proportion to total maintenance*—instead of the strategic objective—*improve reliability of the equipment.* There is nothing inherently wrong with the maintenance tactic that management selected. Improved compliance with a maintenance plan is a good practice that is often associated with improved capacity.

However, as is often the situation, when tactical goals were deployed by management directive and without the context of a strategic intent, the tactic soon became disconnected from its strategic business purpose. Reliability engineers were measured principally on the counted number of programs that they created. Maintenance technicians were measured principally on their success at implementing programmed tasks on schedule. In that environment, reliability engineers did not focus on production-critical equipment and did not spend the detailed engineering time required to develop valuable programs applicable to the most essential work of improving asset availability. Instead, they identified and implemented programs that could be done quickly to increase their count. Often these programs affected the least important equipment and made little contribution to the strategic goal of increasing capacity. Similarly, or consequently, maintenance technicians had no time to practice serious maintenance, but were constantly busy executing the easy and unimportant programs of work that the reliability engineers created.

Because management's tactical mandate to increase program count and compliance had been made absolutely clear and there was no strategic context to highlight the problem, this practice continued for more than 2 years despite the fact that most of our engineers and mechanics knew that it was wrong. When management finally recognized the problem, we changed the tactical emphasis to match the strategic intention. The principal improvement was that we deployed the strategic intent and we changed the way that we measured success. Instead of measuring the *tactical maintenance activity* demonstrated by the quantity of reliability programs created and executed, we measured the *strategic business outcome* of improving equipment performance.

With strategic context and strategically appropriate measures, our engineers continue to develop reliability programs, but now focus their efforts on developing important programs for critical equipment. Craft technicians continue to perform programmed work, but only important work on important equipment. Engineers, craft technicians, and operators share the strategic goal and collaborate to service their unit in ways that enhance performance. We continue to use the original tactic, but we now do it with strategic intent and produce a meaningful business outcome. This experience was a great example for us of the difference in impact between deploying strategies and tactics at the frontline.

Key idea: *When frontline teams know the strategic intent, they can adopt tactics appropriate to that intent.* More importantly, *they can promptly modify or substitute the tactics from time to time, as required to achieve the intent.* When presented with only management-directed tactics, frontline teams often do not have or recognize the authority to modify them. Teams will do what you tell them to do, so be careful what you tell them. One good way to avoid this problem is to be certain that your measures of success are *always* measures of strategic outcomes and *never* measures of tactical activities.

Limiting Opportunities for Improvement

Improving planned maintenance is an example of a tactical goal. As discussed, it may have value in support of a strategic goal but does not have independent value. However, some managers become even less strategic than deploying tactical goals and deploy only a specific tactical action. Again in the extraction area, the first attempt at deploying improvement goals led management to identify the tactical action of using the techniques of single minute exchange of dies (SMED; described in Chapter 4) to speed the replacement of consumable components.

Although accelerating a routine maintenance task is a good way to improve equipment availability, making it the single tactical objective for improvement limits the opportunities of the team. Many other things affect the mechanical availability and performance of equipment and they are all subject to improvement, some with methods other than SMED. Today SMED continues to be used in that area; however, the most important contributions to performance improvement are being derived from close collaboration between maintenance and operations in deploying 5S and operator care (both described in Chapter 9).

Another way in which management can limit improvement is by treating improvement as a single event rather than as ongoing pursuit of a strategic objective. For example, management gave our mine maintenance team the goal of reducing the time required for replacing an engine in a heavy haul truck by 50%. Although that seemed like an ambitious target, the team quickly surpassed that goal and, in the context they had been given, they properly considered themselves to be "done." We later gave them the strategic goal of continuously improving the availability of our heavy haul trucks. In pursuit of that goal, they ultimately reduced the time required to replace an engine by more than 80%. More important, they also made significant improvements to suspensions, transmissions, and other aspects of equipment performance that contributed to keeping the trucks in the mine and productively available.

As you commence policy deployment, you need to communicate the strategic direction of your business. Especially in large organizations, the unifying value of strategic direction is essential to ensure that all of the independent teams work in ways that contribute to the focused needs of the business and that are all mutually compatible. *Absent a clear and shared understanding of the strategic intent of the business, frontline teams can get lost in the implementation.*

DEPLOYING STRATEGIC INTENT

For our purpose, strategic intent is primarily a communication vehicle, so it is important to communicate formally and well. That requires you to deploy *written* goals and to structure the written statement of the goals in a way that enhances its communication value. A good written goal has five separate elements, each of which contributes to communication:

1. Simple and memorable statement of the strategic intent
2. Prose description of the intended future state—once the strategic goal is achieved or has substantially progressed
3. Prose description of the current state—to be used as a comparison in order to identify gaps between it and the desired future state
4. Objective statement of the measurements that will demonstrate progress toward the goal
5. Statement of interim performance targets to be achieved in pursuit of the goal

By structuring your goals in this written format, you can provide all the information needed for people to understand your strategic intent in a convenient and easily displayed form.

Simple Statement of the Goal

The goal statement enables people mentally to carry the goals around with them. It is generally a single sentence, such as "improve the capability and capacity of our plant with investment less than depreciation." When I used that goal at Exxon Baytown, people shortened it further in most conversation to "improve capability and capacity." The reason to provide people with a simple statement of the goal is to enable them to keep the goals in mind and to communicate regularly about the goals. By creating a simple and memorable statement that represents the goal, it becomes possible for people to recall all or most of the important details of the entire goal.

At Suncor, we went one step farther and aligned our goals with four words describing our areas of strategic interest: people, safety, environment, and reliability. Remembering the areas of interest helps people recall the simple goal statements and then the goals themselves. The value of this shared understanding can be so powerful that, with only a single word, we can share an understanding of our common areas of work. For example, we routinely communicate about the competency initiatives (see Chapter 11) with the words "people goal."

Prose Statement of Intended Future State

Following the simple and memorable statement of strategic intent comes a short paragraph or two describing the ways in which the future will be different from the present when we succeed in realizing our goals. This statement should include important objective elements of the intended change. For example, if your people goal is to create a highly competent and engaged workforce, you should describe the future in terms of competence and engagement.

Prose Statement of Current Reality

The next element is a short paragraph or two describing the current state of the business written in the same objective terms as the description of the future. If your goal for the future is competence and engagement, then you should describe the current state of competence and engagement. When taken together, these two statements define the gap that needs to be closed. In effect, you are saying, "We are here and we intend to go there." This gap analysis provides people with a good understanding of your strategy.

Objective Measures of Progress

For people who are still not clear about what you want to achieve after reading your description of the future, what you choose to measure will provide the clarity that they need. For this reason, you need to establish your system of measures carefully because, for many people, the measurements describe what they will do independently of all other attributes of the goal. In the preceding case study, when we began to measure the performance of reliability engineers primarily based upon the number of programs they created, the quality and value of those programs deteriorated rapidly, despite the fact that most engineers understood that this was the wrong outcome.

Interim Performance Targets

Although the reason for communicating the strategic direction is to define the future path of the business over a period of 3–5 years, it is desperately important to set interim targets for performance during that time.

Key idea: *No continuous improvement initiative produces all of the strategic results in the last year of a 5-year plan.* Continuous improvement initiatives should regularly produce measurable interim value. Delayed gratification is unlikely, and the sooner you realize that your actions are not producing the intended results the better you will be able to recover.

I normally prefer to establish interim performance targets for each 6-month period, sometimes sooner. Typically, if you have a 5-year strategic horizon, four 6-month targets in the first 2 years provide a good basis for action. As you progress, interim targets can be refreshed once a year to reflect progress that has occurred and provide a continuously rolling objective for near-term performance.

A 6-month period is sufficient time for a good initiative to produce improvement that is easily distinguishable from normal variation in performance. Six-month targets also lend urgency and provide an objective signal that will help managers or teams promptly recognize if the intended improvement result is not occurring. It is not possible to overemphasize this point. *Continuous improvement at the frontline must start promptly when management creates the capability to practice improvement. Once begun, continuous improvement at the frontline must deliver results immediately and continuously.* During the first several years, the improvement efforts at the frontline will mature and improve, but there should be immediate obvious improvement.

Formatting Goal Statement

Once you have established those five elements for each of your goals, I recommend communicating them formally by using one side of a single sheet of paper containing each of the five elements for each goal. In that way, it is relatively easy to post your organizational goals as well as the goals that others throughout the organization derive in response within their own work areas. Thus, you can make the goals of your improvement process visually apparent in many places. Nothing facilitates valuable local discussion and understanding more than having the original corporate statement of strategic intent as well as the local strategic response hanging on the wall for everyone to see. Figure 3.3 is an example of a goal prepared in that format.

TRANSLATING STRATEGIC INTENT THROUGHOUT THE ORGANIZATION

The second element of policy deployment is translating the strategic direction throughout the business. Once you have decided upon the goals that frontline teams can help you achieve, you need to get the word out. These goals are not for managers and they are definitely not for corporate plans that are prepared annually and then filed away. These goals are only valuable when they are routinely used in a way that enables people throughout the business to help you achieve them. That happens when you communicate the goals so that each goal *belongs personally* to each middle manager—and, ultimately, to each frontline team—in terms that are meaningful to those individuals and with measures meaningful to their work.

The most critical aspect of this work, beyond ensuring that your business goals accurately arrive at the frontline with locally meaningful interpretation, is making certain that you proceed through the organization and do not attempt to go around the middle of the management structure directly to the frontline. No matter how sincerely or urgently senior management wants help from the frontline, the frontline must receive guidance from the middle managers from whom it normally receives work direction.

If the middle managers are not involved personally in creating the new system that will be used for autonomous improvement, it will not succeed. Eventually, senior management must turn its detailed attention to other things or the practice of autonomous improvement will grow so large that detailed senior management attention in all places is impossible. When that happens, the initiative will have no support at all unless the middle managers who normally lead the details of your business have been personally engaged in creating and leading your new practices.

Furthermore, by carefully translating the goals through the entire organization, there are multiple opportunities to ensure that the derivative

Strategic Goal

To maintain our license to operate.

Vision of the Future

Our business needs to demonstrate that our operations are absolutely legal, safe, honest, and ethical. We need to be recognized as good neighbors. By year-end 2010, we will meet or exceed all then-current legal and regulatory requirements for safety, health, financial and audit controls, and personnel standards. We will have in place a process to anticipate future requirements and this will be a primary input to our operations and business planning. We will improve personal relations with our neighbors and the local civic entities to become a trusted and valued member of the community.

Today's Reality

We are required to comply with various standards mandated by law, governmental regulations, industry associations, financial accounting standards, or corporate policies. Our method for achieving compliance is often reactive and we do not achieve compliance as well as or as effectively as we would with adequate anticipation of new requirements.

Community involvement is excellent and community support is adequate. Workplace safety is far better than industry average, but not as good as we want it to be. The effectiveness of our efforts in wellness initiatives is not measured and not well known. We are in compliance with all applicable laws and regulations. Our current compliance with accounting standards and corporate policy is not satisfactory but is improving.

Measurements

Monthly

- total recordable incident rate for all people including both employees and contractors
- count of community complaints
- count of regulatory actions

Quarterly

- total air, water, and land environmental emissions and trends

Annually

- safety and wellness program assessments using standard assessment format
- financial and internal controls audits using standard audit format
- count of governmental actions or complaints and measured regulatory action
- legislative advocacy program development in cooperation with industry association

Performance Targets

- total recordable incident rate equivalent to or better than average of top five U.S. industrial companies by 2010 with ratable progress toward that achieved during each 6-month period
- community complaints reduced by 90% by 2010 with ratable progress each 6-month period
- total environmental emissions in tons reduced 50% by 2010 with ratable progress each 6-month period

FIGURE 3.3
A properly formatted goal statement.

goals and actions throughout the enterprise add focused value to the common objectives and that all goals are compatible with one another. In large organizations—especially in organizations with independent staff or functional areas—it is far easier than you can imagine for the teams to adopt inconsistent goals unless you formally take great care to ensure that this does not happen.

Support and consistency are achieved by following what has come to be called the *three-level view process.* As Figure 3.4 shows, the three-level view consists of a simple overlapping and stepped process for reviewing the goals and actions of the organization in relationship to the source of the goals, the goals and actions that are shared with colleagues, and the goals and actions that will be passed on to others. As the figure illustrates:

- At the highest level, the executive committee (EC) establishes the goals for the corporation and shares them through the executive vice president (EVP; who is also a member of the EC) who leads each independent business. Then the EVP for each business shares the goals with his or her business management committee (MC). This is the first three-level view: EC as a team, the several business EVPs individually, and the business MC reporting to each EVP.
- At the next level the EVP for each business works with the MC for that business. These MCs comprise the several vice presidents (VPs) of operating units within each business. In collaboration with one another and with the EVP for their business, the VPs act as a management committee to derive the goals for the entire business from the goals they received from the corporate EC. They also decide which goals each will take forward into his or her operating unit. The management team within an operating unit comprises the VP and the directors who report to the VP, so it is called the management committee of directors (MCD). The second three-level view is the business EVP, all the VPs together, and each VP with his or her MCD.
- The VPs, each within his or her operating unit, translate the goals to their management team (MCD). Each member of the MCD translates the goals to the managers in his or her department. This is the third three-level view: VP, the MCD, and the managers reporting to each MCD member.
- At the next level, each of the directors or general managers (GMs) works with his or her managers (direct reports) and translates the goals in the same manner; the managers carry the goals to the

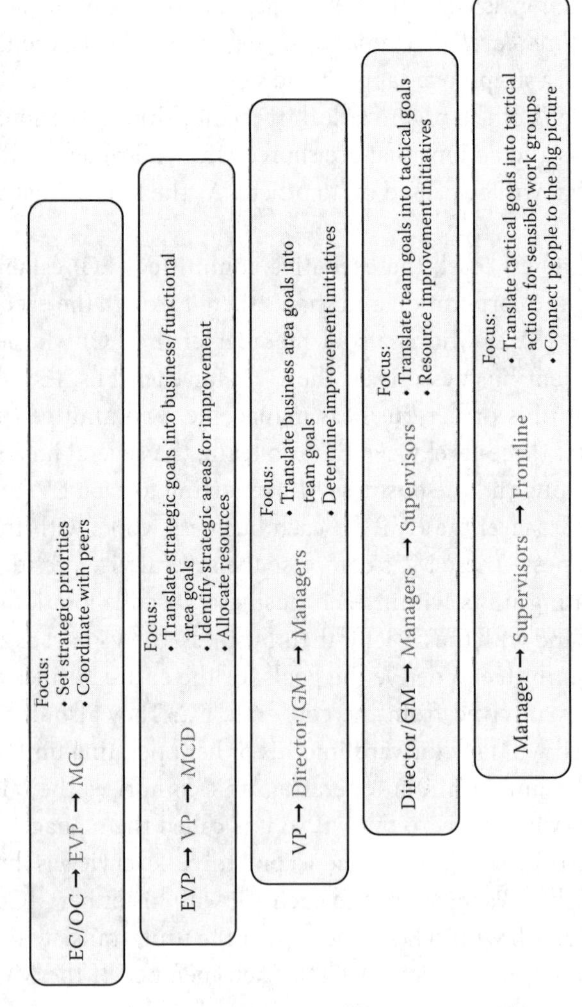

FIGURE 3.4
One form of three-level-view process.

supervisors who report to the managers. The fourth three-level view is the MCD, managers with their teams, and the supervisors reporting to the managers.
- Finally, the managers work individually with their team of supervisors to complete the translation of corporate goals into a form that is meaningful to each of the frontline teams. The fifth three-level view is the manager, supervisors, and each frontline team.

Within Suncor, at each level, the teams are also cross-functional teams; therefore, in addition to direct business reports, each line manager's team includes representatives of the supporting functions, such as maintenance, supply chain, reliability engineering, and others. In this cross-functional relationship, some members of the team arrive with delegated goals from a function and some with delegated goals from a business. But all goals are ultimately derived from the corporate goals and the cross-functional teams greatly benefit from the new source of consistency of purpose.

In this manner, everyone in turn is part of the process. Each participant recognizes the source of the goals, the actions that will occur at his or her own level in a collaboration among peers, and the resulting goals that each will carry forward to the next level. Because each middle manager participates in taking the goals forward into the organization, the goals of the business do not arrive at the frontline as unintelligible words from senior management that have no meaning at the frontline. Goals do not arrive at the cross-functional teams as a confused and inconsistent mix of diverse objectives originating in multiple ways. Rather, the goals arrive as a thoroughly understood and agreed-to set of objectives that exactly matches the work at hand and the expectations of the immediate supervisors who will lead the work. These are goals that people can achieve and, generally, goals that will make their lives or at least their work better in the process.

To illustrate how simple and yet powerful this deployment is, let us look again quickly at the memorable goal statements for the several translations of Suncor's goal for increased output as it progressed from the corporate offices to the cyclonic separator maintenance team:

- Suncor executive committee: *Increase production from existing plant and equipment.*
- Oil sands EVP: *Increase production of synthetic crude oil.*
- Extraction VP: *Increase production of bitumen.*

- Director of cyclonic separation: *Increase bitumen separation.*
- Manager of maintenance and reliability: *Increase availability and reliability of cyclones.*
- Supervisor of preventive maintenance: *Increase effectiveness of routine maintenance and decrease unproductive time consumed for routine service.*

That progression may not seem very significant when it is presented as a completed set of translated goals. However, when coupled with the other detailed elements of a good goal, the result at the frontline is that people actually receive a strategic goal they can act upon with the real certainty that they are contributing to the business in a way that adds value and is compatible with other teams' efforts.

Consider the previous example describing the tactical goal of increasing planned maintenance as a proportion of total maintenance for the unit. Because maintenance management did not maintain the simple link to the strategic direction of the business, it chose to adopt a maintenance tactic as an improvement goal and did it in a way that made business performance worse rather than better. Essentially the same thing happened to the supply chain organization described in Chapter 2 that adopted a "lean inventories" goal in a way that made maintenance and operations worse rather than better.

In both cases, honest, hardworking managers intending to improve their *functional performance* lost the tactical connection with the strategic improvement goal for *business performance.* Countless examples demonstrate the proposition that *without a formal effort to ensure strategic alignment, some teams will surely get off course.* In a traditional management environment, that happens often enough that you can easily imagine how frequently it would happen in an environment of autonomous action at the frontline without a clear process for ensuring alignment.

As we finished the goal deployment discussions at the management committee for Suncor's oil sands business, one of our vice presidents observed, "Last year [our EVP] had 15 goals and we spent 4 minutes talking about them. This year he had four goals and we spent 15 hours talking about them with more to come." That is a great description of the process and a valuable lesson. As we set the goals of our business, we carefully examined the goals we had received from the corporation. As leaders of seemingly independent units within the business, we spent the time required to understand how we worked together to achieve those goals. We also shared our intent for specific work within our separate parts of the business.

Whereas we had previously acted much like individuals who periodically sat together at the same table, the work of translating our goals transformed us into a true team with shared objectives to which we all contribute. An important element of goal translation is the new collaboration among different teams and individuals who recognize through goal translation that they are indeed pursuing a common goal through shared efforts rather than pursuing independent goals through parallel efforts. By spending the relatively small amount of time needed to achieve that shared understanding, all of our efforts for the past year have been much more collaborative and successful.

Key idea: *The true value of goals* is not in possessing goals. The true value *is transforming the goals into shared understanding of mutually supportive actions that will advance the business.*

FRAMEWORK FOR ACTION

The final element of policy deployment is the creation of a framework for action. Even after people at the frontline have clear strategic goals that they can understand and execute, they still need a formal way to demonstrate that each of the tactical actions they will take in pursuit of the goals is consistent with the intent of doing the "right thing in the right way." The framework for action is the place where the special effort necessary for careful strategic deployment within the process industries becomes especially apparent.

Key idea: The strategic direction describes the *right things* to do. The framework for action provides the *boundaries or limits* within which people will have authority to act.

All autonomous industrial teams benefit from the focus and communication provided by a formal practice for making their improvement efforts visually apparent. In the process industries, there is special benefit to this practice. We not only need to ensure that all efforts add value and are mutually compatible but also that no team ever makes a change that introduces or activates a new risk of injury or environmental damage. Ours is a serious business that demands close attention to detail, and attention to the details of creating and managing change becomes especially important as we engage people who have not previously had any authority to make changes. As will be described, the framework for action and the rules of practice that accompany the framework provide the management oversight and assurance required to allow people to practice autonomous improvement. *Autonomous improvement is not closely supervised, but it is always very carefully managed.*

As the teams commence autonomous improvement, they need a way to track progress that demonstrates that their work has achieved value. They also need a way to communicate internally among the team members to propose, evaluate, and select new projects. Management needs a method that allows it to monitor progress and success as well as a means to ensure compliance with the limits on the authority and actions of teams while not intervening more than required in the activities of the team.

How Quality Stations Work

The framework for action provides all that and more. At both Exxon and Suncor, we call these frameworks for action "quality stations." A quality station must do four things to meet these needs:

1. It must demonstrate the goals the team has received and what they intend to achieve locally in response.
2. It must demonstrate through objective measures of the goal what the team has previously accomplished.
3. It must show what is now in progress and what the objectively measurable outcomes of that work will be.
4. Finally, it must have the capability for people to interactively propose, evaluate, and select new work.

Physically, a framework for action can be whatever the team wants it to be so long as each one has all four elements that will enable it to serve its purpose. I have seen quality stations prepared with great attention to detail and almost beautiful in execution and I have seen them composed

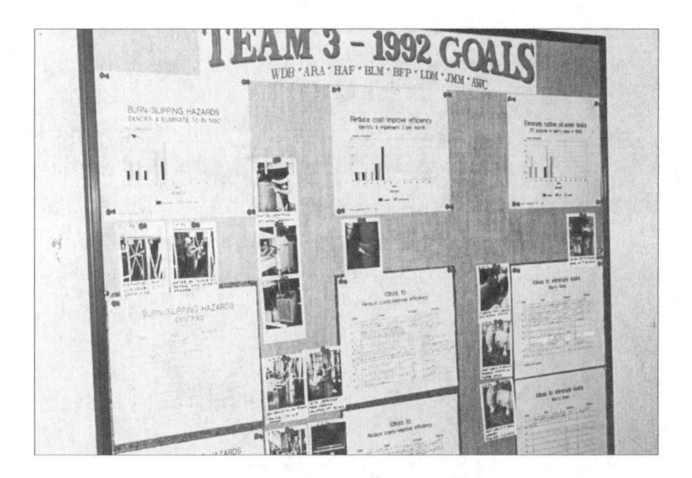

FIGURE 3.5
A quality station at Exxon Baytown.

of nothing more than a bunch of yellow-pad sheets stapled to a corkboard. The appearance of a quality station is only significant in that the team that owns it must be willing to maintain it. Often I find that because of that requirement or the lack of it, the frameworks soon begin to look very distinctively like the personality of the team.

Because quality stations have no standard format, I always hesitate to show samples, but I am always asked to do so. Figure 3.5 is a quality station from Exxon Baytown and the one in Figure 3.6 comes from Suncor. Figure 3.7 is a quality station that I found in Japan. Even in the land made famous for standardization, the form used by individual teams to display their work can be very personal.

At the most fundamental level, a quality station is a place where the team members can gather to discuss and direct their shared work and communicate about their work with others. It is also a way to make the work of the team visible so that management can exercise both business and safety oversight with little or no intervention into the work of the team.

Display the Team Goals

Each frontline team needs strategic goals that are specific to the work of the team. Those goals *need to be obviously derived from the corporate goals and they need to be visibly displayed.* When a team begins to function well in the autonomous improvement mode, it is normal for a good team to average 40 or more improvements per person each year. For a team of

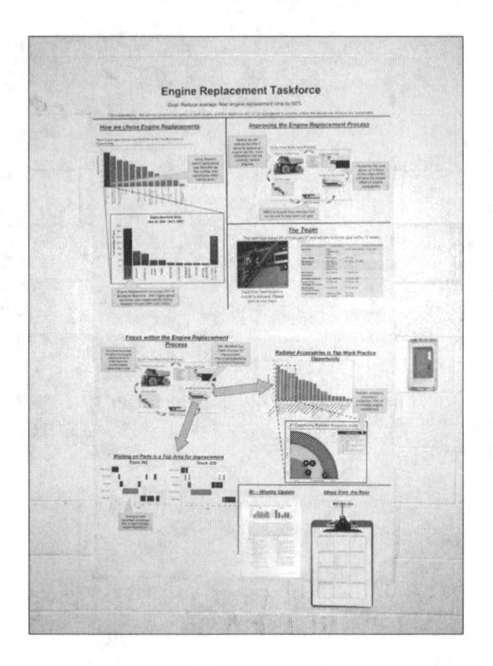

FIGURE 3.6
A quality station at Suncor.

FIGURE 3.7
A quality station in Japan.

10 individuals, that rate of improvement equals 400 individual improvements over the course of a year.

In such an environment of rapid change, it is very easy for a team to lose its focus and begin to drift away from the original strategic intent incrementally. Requiring that each team continuously display its goals and the derivation of those goals is a constant and visible reminder to the team members and others of what they are supposed to do. Each completed project can be reviewed to ensure that it supported the goals. Each new project can be selected so that it will support the goals. Should any one of the projects in progress begin to develop in a way that does not support the goals or if it appears that it might detract from the goals, it can be modified or abandoned before harm is done.

Quality stations control the goals as they are deployed and require that each team constantly and visibly demonstrates that all of its actions are continuously aligned with the goals and do not exceed the limits on the goals in any way. Quality stations are the first of several forms of noninvasive oversight that enable autonomous improvement at the same time as they maintain process integrity. If it is always visibly apparent that the teams are doing the right thing, management can generally allow them to do it.

Display What the Team Has Completed

For several reasons, each team must display the objectively measured results of the work that it has completed. First, because the whole process exists only to make improvements, teams need to demonstrate that they are making goal-aligned improvement and doing it at a good pace. Again, this is an opportunity for noninvasive oversight. If the teams are making good progress, management does not need to intervene in their work. Any time that a team is not able to demonstrate that it is making substantial progress, management does need to intervene in order to provide whatever help or direction the team needs in order to start or restart the improvement process.

Second, the objectively measured outcome of the projects that a team has completed is the best indicator that its members understand their goals in the same way that management and others do. Objective measures that are selected to track progress are often more powerful communicators of intent than words in defining goals. As a result, many teams focus in surprisingly *inappropriate* ways on the metrics rather than on the prose description of the intended future state. Even when a team is performing well against its measures, it is valuable periodically to assess the outcomes against the intentions independently. This review is a further assurance that the work of the team is within the limits that management has established for autonomous improvement.

The third reason for requiring teams to demonstrate the results of their work is that as this practice develops in your plant, you will find that the work of successful teams will become a good source of ideas for other teams that need some inspiration. Even good teams often get inspiration from the work of others.

Show the Work in Progress

Each team needs to have at least one improvement project in progress at all times and needs to demonstrate the progress of that work on its quality station in an easily understood manner. What the members of the team are doing and what they anticipate achieving as a result should be apparent. Teams that are a little more advanced frequently use something a bit like a formal project management approach and include estimated spending, forecast completion dates, and other interesting data.

However, the important element is that teams must have work in progress and management must be able to monitor this work to ensure that teams are making progress focused on the goals and properly within the boundaries. If a team is not making improvements, management must ensure that it receives the help it needs in order to do so.

Note: Just as it is important for managers and engineers to initiate the autonomous improvement process by agreeing to the goals and boundaries for the frontline teams, it is equally important for managers and engineers to review continuously what has been done and what is being done in order to maintain ongoing oversight of what the team is doing.

Provide Interactive Space

An interactive section where team members and others can propose new ideas for future team action facilitates two primary functions. The first is the essential element of proposing new ideas for team members to consider. If the team is to work on a continuing series of improvements, it must have a continuing series of proposals from which to select those projects.

The second attribute is that the interactive portion of the quality station is where managers and engineers get a very formal and specific chance to review proposals before the teams begin work. In process industry applications, these interactive sections normally have very specific "rules of practice" that are directly aligned with the plant's management of change policy. These rules will likely include a requirement that any new idea must remain as a proposal for at least 2 weeks before it is promoted into autonomous action. The reason for imposing this delay is to allow time for others, including managers and engineers, to look at proposals in a "cold eyes review" prior to allowing the team to commence work. This allowance for a managerial and technical review is a limit on the authority of a team to act autonomously, rather than an absolute limit on the improvement process. If a team finds a project it wants or needs to implement promptly, it can do so if it asks for and receives an ad hoc technical review to comply with the traditional formalities of change management.

This review of work before it commences provides still further assurance that the actions of the teams will always be the right things done in the right way and within the established limits. It also ensures that, if management discovers a proposal that requires intervention, the intervention will occur *before* the team becomes invested in the project or before any actions have been taken that might need to be undone or redone. In this way, managers can maintain business control and engineers can maintain technical control without generally intervening in the team's activities.

Other rules of practice are developed as appropriate. For example, we never allow any autonomous action at any time that touches the hydrocarbon containment envelope or the pressure containment envelope of our processes. Still other rules of practice are local: For example, each team will develop its own process for assigning work, selecting projects to promote from idea to action, or the process for allocating resources such as budget and time.

Policy Deployment in Action: Conversations at a Quality Station

Policy deployment is principally an exercise in formal and precise communication throughout an organization. The quality station is the final element in that chain. Two important parts of the communication exercise take place at a quality station. First, the frontline team members communicate internally about their goals, their work, and their results. Second, team members use the quality station to communicate with other people. For many in the frontline of your business, these will be the first conversations of this sort in their careers. As a result of these conversations, they will uniformly begin to engage with your business in a completely new way.

Frontline people routinely tell me that becoming an informed participant and a trusted and valued colleague in making the business better is their most rewarding work experience. At Exxon Baytown, an electrician who had been in our plant for 40 years when we began to engage people in the improvement process told me that, for the first time in his long career, he had the sort of job that he had hoped to get when he joined the company!

A similar thing just happened at Suncor. An executive of our union who was formerly a craft technician asked to attend our SMED training (see Chapter 4). Although SMED clearly has nothing to do with his union role, as a craftsman he felt that he wanted to participate. Like the Exxon electrician, he said that we were now doing the sort of thing that he had always wanted to do.

Internal Team Conversations

Policy deployment in the field begins to have practical effect during the formative discussions as the team receives its goals through the translation process. During this activity, most frontline teams discuss for the first time how they can contribute to the success of the business in ways beyond the execution of their routine tasks. That initial discussion often is more theoretical than useful until the team has some experience with frontline improvement. After the team members gain some experience, it is common for teams to revisit their goals and produce better strategic direction as well as better tactical action. That process of constantly improving the goals of the team is perfectly appropriate as long as the goals receive management review at all stages to ensure alignment and boundaries are preserved.

Teams find that once they have established a good strategic direction, their objectives remain fairly constant but their tactical goals and the specific actions that support that strategic intent evolve very fast. This is an exciting time for the teams and they will engage each other in a lot of conversation about the goals and the tactics. That exchange of information and opinion enhances their knowledge of the unit and the process.

Everyone who has ever worked in industry has experienced some time when management has asked the frontline for improvement initiatives. But this time, particularly when combined with the leadership activities described in Chapter 10 and the tools of autonomous improvement described in Chapters 4 through 9, this request begins to look and feel as if it will succeed. Management is not just asking people to provide improvement. Management is obviously doing things that will let it happen and make it real. That is very different and employees promptly recognize and appreciate the difference.

Key idea: An old maxim says that it is better to seek forgiveness than to ask for permission. The implication in industry is that it is acceptable or even encouraged for people simply to begin making improvements—to “just do it.” Managers often adopt that approach when they seek frontline improvement in an unstructured way. However, for most people at the frontline, seeking forgiveness in the event something goes wrong is not a good bet. Equally, for managers in the process industries, authorizing unmanaged change is not a good bet. If we expect to receive the benefit of frontline improvement, we must create a carefully managed framework for action within which everyone can be certain that people are doing the right things in the right way and staying within the boundaries.

Throughout goal translation and continuing into actual improvement practice, people begin to talk about their work in light of the strategic goals of the business and about how they can do things within their own capabilities that will contribute to the goals. In the same way that management committees discuss the goals and actions that will improve the business, members of small teams do the same. At Suncor, we gave each team 8 hours spread over several shifts to start this process. At the end of that time, the teams could say with some confidence that they knew how they could change their work to contribute to the goals of the business. After that good start, they continued the conversation as part of their regular team meetings.

Before the small teams can engage to make changes, they need the full suite of cultural enablers (see Chapter 10), but policy deployment that enables people to understand their contributions fully is a powerful first step. Even before they are able to commence autonomous improvement, you and they will benefit from understanding their unique role in the success of the business.

Once the teams understand how they can contribute to the business, the best conversations within a team start with an idea for future team action. A member makes the proposal by putting it in writing on the interactive portion of the quality station for team members and others to review. This is far different from sending a suggestion to the engineering department for consideration because the person is proposing an idea about the immediate work to the people with whom that work is shared. Whether or not the individual is gifted at writing proposals has little impact on the outcome (unlike a traditional suggestion program). The people who will be reviewing any proposal are already very familiar with the operating context, probably very familiar with the problem to be solved, and they work every day with the person who made the proposal. Any uncertainties are easily managed.

I have always believed that a key factor in the failure of standard suggestion programs is that you will never receive a person's second good idea until you have done something he or she respects with his or her first good idea. Such a happy outcome is rare in traditional suggestion programs, but that outcome is regularly achieved at quality stations, where ideas are reviewed in detail by the folks most familiar with the specific situation.

An extremely interesting aspect of proposing improvements to team members who share the work is that ideas often mature or evolve from one idea into another, far better idea *before* any work is done toward implementation. I have seen very modest original concepts develop into truly grand projects in this way. In one interesting example, an original proposal to change the safety guard on a building exhaust system transformed into a complete redesign of the mechanical elements of the system.

This extensive change was truly autonomous work despite its scope because the work could be done incrementally within the capabilities of the maintenance team that proposed and undertook the changes. There was no real incremental cost to implement the idea. Each element of the system was converted to the new design when it needed maintenance. In that way, the cost of conversion was immediately offset by avoiding the cost of repair. The result was the transformation of an exhaust system that originally required extensive and expensive professional maintenance into a system that the unit operators could maintain easily and inexpensively. The final project resulting from the original idea incorporated more than 30 additional and different ideas from various team members.

In addition to understanding the goals and deciding how they can contribute, small teams at the frontline conduct other conversations. Teams normally have a small budget (see Chapter 10) for implementing their changes and they need to decide among themselves how to spend that money. They also decide how they select proposals for new work and how to allocate work among the team members, which often implies that someone is working on improvements while others are running the unit. The transformation is amazing. People who previously had little engagement promptly begin to take a very proprietary approach to all aspects of their operation. Frontline teams begin to look and feel a lot like small businesses, each very successfully running its part of your plant.

External Team Conversations

One of the more socially interesting changes that occurs when teams adopt a quality station is the conversations that team members have with people who are not on the team. Traditionally, at the frontline, substantive talk with people who are not immediately involved in the work at hand is rare. The job of most people in a traditional environment consists of routine work and only modest change, which progresses at a slow pace. In addition, all of the change has been initiated by others. The team members have no role in the change except to experience it either gladly or grudgingly.

When senior management or other visitors come to the work area, the conversations are almost exclusively focused on safety or are social in nature and very superficial. Senior managers spend a lot of time talking with people about sports and the weather. When the conversation does turn to business, it is often either the manager quizzing the employees or the employees complaining to the manager. Neither conversation is very satisfactory to anyone. The result is that many really good people do not choose to talk with senior managers and other visitors to the work area. They simply turn and walk away when they see someone coming. Each time that happens, a great opportunity for the business has been lost.

A quality station changes all that. A lot is going on and the people at the frontline are deeply engaged in making it happen. The team conceived the changes in progress and they have implemented the completed projects. By the way, in addition to what has already been done, more improvement is in progress right now and more ideas for further improvement are coming behind that. Once a quality station team gets well underway, there are always rich conversations with visitors.

For example, a very senior executive from Chevron visited us in Baytown one day with a team of about 10 senior executives from the refining organization. At the end of the day, he was very complimentary of the folks he had met and observed that I was very fortunate to have such a fine group of executives and managers. What a surprise he and his colleagues had when I told him that I was the only person on his agenda for the day who was not a member of the union.

As we experienced this transformation at Exxon Baytown, I began to have the most interesting conversations with people who were literally waiting for me to come to their work area so that they could talk about the things they were doing. During this period, we routinely held a portion of each management committee meeting in the field and the opportunity to represent a quality station team to the management committee became a desired form of recognition.

Visiting the quality stations of other teams also represents a powerful opportunity for change in the way that people work and in their approach to work. In this way, good ideas move around the plant and improvements are cross-pollinated and transformed into better ideas in other places.

Case Study: The Value of Talk

This effect is especially powerful when the teams involved have some sort of relationship. For example, in the nature of our operations, Suncor's upgrading and extraction units are heavily dependent upon water volume and water quality. Historically, water has been managed for the site by the extraction organization. When the upgrading operators at Suncor were able to understand the impact on the performance of their operations that resulted from water quality as it arrived from extraction and the extraction operators were able to understand the impact on their operations of water volume as it returned from the upgrader, they began to collaborate in truly remarkable ways, which have provided a great benefit to the business. That discussion and that outcome were both possible previously, but they had not occurred previously because it was never before clear to people at the frontline that the two units shared common needs and objectives.

In combination, *lean enterprise thinking* and *policy deployment* will create the essential knowledge of strategic intent required to enable the practice of lean improvement at the frontline. People will be able to identify the *right things* to do.

In the next six chapters, we will examine the essential tools of lean practice as they apply to the process industries. As you master each tool, you can promptly deliver it to your teams. They will not yet be ready for truly autonomous actions, but they can immediately begin to practice improvement with specific management or supervisory guidance and support. In this way, they will gain expertise and experience with the new tools that will later become the basis for autonomously doing the right things in the *right way*.

4

Improving Flexibility and Availability in Mechanical Equipment

INTRODUCTION

Lean values require that production equipment must avoid the wastes of production outages (availability), ineffective product transitions (flexibility), and poor product quality (capability). *In ideal lean production, a plant is able to make any product at any time in any quantity.* Of course, the product itself should always be good.

In liquid manufacturing, much of the burden for achieving those high expectations falls upon the equipment. Indeed, because liquid manufacturing is so capital intensive, one of the greatest opportunities to improve process operations is to improve the capability, flexibility, and availability of the equipment used. (*Capability* is the subject of Chapters 7 and 8, *flexibility* and *availability* are discussed here, and *availability* is also covered in Chapter 9.)

One of the most useful tools of lean, known as single minute exchange of dies (SMED), is especially applicable to problems of flexibility and availability. Whether the production loss occurs because equipment requires reconfiguring to make a new product (flexibility) or simply needs repair or maintenance (availability), SMED can greatly reduce the losses associated with any mechanical task. In practice, the two uses of SMED are almost identical, so we will not differentiate between the two in this description of the technology.

Note: Even if you operate a petroleum refinery (as Suncor does) or similar unit that genuinely does not need to improve its flexibility, you will find

SMED a useful tool for improving other mechanical activities that affect availability, including routine maintenance.

SINGLE MINUTE EXCHANGE OF DIES SYSTEM

The SMED technology lies at the heart of improved mechanical flexibility and improved routine maintenance. It was first described in the book, *A Revolution in Manufacturing: The SMED System,* by Shigeo Shingo (Productivity Press) in 1985. I got my first copy shortly thereafter and immediately fell in love with it. Since then I have routinely bought copies of Shingo's book by the case and handed them out to colleagues who I believed could put the concepts to good use. I encourage you to get a copy and put it to good use.

Because SMED is an acronym for "single minute exchange of dies," my colleagues often remind me that in the liquid process industries, we do not have dies. Of course, we do. Densification of polymer granules (discussed later) uses pelletizing dies that have a function very similar to the forming dies used in the industries that originally gave SMED its name. That said, successful use of SMED in the process industries requires some translation in both language and practice and we will do that here.

Case Study: Reduce Maintenance Time with SMED

We first introduced SMED at Suncor in the mine maintenance team that provides routine service for our heavy haul trucks. In their initial use of SMED, they reduced the time required for an oil change on a 400-ton truck from 83 minutes to 8 minutes. In their second use, they reduced the time required to change a truck engine from 250 clock hours to less than 45 hours. With those two experiences as a good start, within the past 8 months the mine maintenance team has improved truck availability by more than 55% (relative) from prior levels. As described in some detail later, our experience in replacing heat exchanger bundles as well as many other opportunities for improving process equipment has been similarly rewarding.

In this chapter, a continuing series of examples is based on a plastic extruder at the Exxon Mobil Baytown chemical plant. To begin, the following case study is a brief introduction to that operation.

Case Study: Reduce Manufacturing Time with SMED

Most manufacturers who produce plastic products feed their extruders and injection molding machines with dense plastic pellets because the pellets are convenient to handle and are easily fed into processing machines. However, plastic, as it emerges from the polymerization reactors, is usually in the form of low-density crystalline granules. In the plants that produce polymers, those granules are "finished" by using very large extruding machines to make the crystals denser and turn them into the pellets that are used by our customers.

Because pelletizing is the final manufacturing step before the product reaches the customer, it is also the step used to introduce customer-specific additives or modifiers to enhance or distinguish the product. Changing an additive often requires physical modification of the extruder or the support equipment, which can be a modest change such as switching the material in the additive feeder or a much larger change like reconfiguring the extruder barrel from plugged operation to vented operation.

Because the plastic cannot stop in the hot extruder, in the past, the reconfiguration required that we terminate production, purge all the material from the extruder, cool the machine, reconfigure the equipment, reheat the machine, and restart production. Often this activity required more than 15 hours of elapsed time from the moment that production stopped until the time that production resumed.

Using SMED, we reduced the actual work time required to execute the mechanical reconfiguration to just a few minutes. With such rapid execution of the work, we can benefit from the persistence of the material as it flows through the machine, so we can now reconfigure the extruder while it remains in relatively normal operation. Today, we routinely run continuously throughout this mechanical conversion with only a very small loss (less than 100 pounds) of transition product at the interface that occurs when we engage the new additive.

Key idea: A good result in mechanical production is to reduce the manufacturing time lost during equipment reconfiguration from 4 hours to 10 minutes. A good result in process manufacturing is to reduce the production loss from 15 hours to effectively zero. *Because of considerations unique to the process industries, our initial problem with flexibility is much greater than it is for a mechanical manufacturer, but our opportunity for improvement is also much greater.*

As you will see, in reconfiguring this extruder, we are essentially performing the mechanical operation of exchanging one part for another.

Although in this case the work is done to increase flexibility, the lessons from this example also describe the practice of improving availability by increasing the speed of routine maintenance.

What We Can Learn from NASCAR

It is useful to begin the discussion of SMED with an analogy to a situation that is common to everyone: a flat tire. If you go to the parking lot and find that your car has a flat tire, you know that you are about to lose a half hour or perhaps more. By the time you resign yourself to the situation, get out the spare tire, figure out how the jack works, get out the manual to identify where the jack attaches to the car, and find the tools, you might lose 10 minutes or more before you begin working on the tire itself. Unusual complications aside, it takes about 30 minutes to change a tire.

Yet we know that every Sunday afternoon we can turn on the television and watch the professional NASCAR pit crews change four tires and fill the car with fuel—all in 12 seconds. What are the characteristics of race cars and their pit crews that make this dramatic difference in performance possible? More importantly, can you duplicate or adapt those characteristics for your purposes? The answer to these questions is that racing teams have configured their equipment and their work to "keep the cars on the track"; by using SMED, you can also configure your equipment and work to keep your plant in service.

Actually, the differences between NASCAR pit crews and the maintenance teams in your plant are not as great as you might think. Professional pit crews have three primary advantages that the stranded motorist, an amateur, does not: preparation, teamwork, and equipment. Because we are professionals at our work (operating process plants), we should be able to adapt their professional advantages for our use. As you will see, these NASCAR practices are essentially the same as SMED practices:

1. *Preparation:* NASCAR teams are fully prepared. All the tools, equipment, and parts that they need are staged at hand ready for the event. They also have contingency tools and materials staged at hand in the event that some part of their preparation is imperfect or something unexpected occurs.
2. *Teamwork and practice:* NASCAR teams have several people, all of whom are well practiced at working as a team. In that way, they can

effectively perform the separate elements of the job simultaneously as opposed to having a single person doing each task in sequence.

3. *Equipment:* NASCAR teams recognize the value of reconfiguring their equipment faster than the competition does, so they have spent the time and money needed to modify the vehicle, the wheels, the team, and the tools or equipment to make the task easy and fast.

The third element contains an essential concept. NASCAR teams have recognized the value of keeping their equipment in service. Although I stopped tracking this statistic some time ago, for a long period, the Indianapolis 500 race was not won by the team with the fastest car on the track; it was won during the pit stops. That is, the total margin of victory in the race was less than the marginal time difference in servicing the cars.

Case Study: The Value of Keeping the Cars on the Track

In 2008, Ryan Newman won the Daytona 500 by a margin of only 0.092 seconds. During an interview that appeared in the May 9, 2008, edition of the *Wall Street Journal*, Mr. Newman's chief tire carrier said, "You can't win a race with a 12-second stop, but you can lose it with an 18-second stop." Pretty much the same is true for process manufacturers.

Key idea: Rapid execution of mechanical work alone will not make us successful, but poor execution can make us unsuccessful.

Translating NASCAR Success to Our Plants

Let us look quickly at the details of each of the NASCAR advantages, which should help you see the industrial applications of this technology in your business.

Preparation

Not surprisingly, simply being prepared to do work has enormous advantages. Part of that preparation is establishing close coordination so that the crew is ready to work at the first moment at which the equipment stops. Part of that preparation is organizing the work and the workplace by having the appropriate support equipment, parts, and tools standing by. While the race car is stopped, *no activity occurs that is not absolutely and immediately related to and necessary for returning it to operation.* Every other activity is done either before the car stops or after the car restarts. We can adopt the same intense focus when the equipment stops in our plants.

Key idea: Preparation activities do not cost more when done prior to stopping production. Often they cost less because they can be done in a more orderly manner. The only important difference between preparing for the work before it starts and preparing after it has started is that the plant continues to earn money during the period of preparation.

Case Study: Make Gains by Paying Attention

Suncor's first major increment of SMED improvement—the first 100-clock hour reduction in the time required for changing an engine—occurred before we began truly to practice the disciplined technology of SMED. While we were gathering the details of the work in order to perform the SMED analysis (described later), we changed an engine in half the prior elapsed time. It happened because during this event everyone was paying attention to the work. Until then, unlike the NASCAR teams, our mechanics had not yet recognized the value of keeping the truck in the mine, or at least they had not yet applied that value to their work.

There is no reason why we in the process industries could not experience a NASCAR level of preparation for every event where we need to reconfigure our equipment and for all of the maintenance events that stop production.

Key idea: As with the Suncor trucks, you can likely receive an immediate incremental improvement from that type of focus simply by making the inherent value of routine work more apparent and by giving your teams the opportunity to respond to that value proposition. Remember that the quote earlier on the value of the pit stop did not come from the driver or even from the crew chief. It came from a tire carrier.

The principal difficulty in promptly achieving NASCAR-type preparation in our plants is that the volume of detailed planning required to apply this approach uniformly to all of our equipment exceeds the capabilities of our professional planning staff. For that reason, SMED is best applied as a frontline practice where people in small teams can plan and organize their work. Small teams in our plants have the same size and natural focus as the NASCAR crews. In teams of that size, people can plan and execute their work at a level of detail that can never be achieved by engineers or even professional planners.

Because it enables people throughout the business to achieve highly detailed improvements at a level of detail the engineering staff could not attempt, SMED is one of the great tools of autonomous improvement. There is essentially no risk in authorizing autonomous SMED practice because the value stream of the work and the work product are generally not changed. As you will see, *the principal change is to eliminate waste from the execution of the work.*

Note: Although SMED is thought of as a tool for deployment to the frontline, if you have a need or an opportunity to do so, you can devote professional staff to this work as soon as you complete this chapter. The success of that effort will be an important example to the rest of the organization.

Case Study: Stage Materials and Reminding Teams about Procedures

The maintenance manager at Exxon Chemical Baytown recognized an opportunity and created Exxon's first SMED improvement less than 2 hours after his initial introduction to the technology. He applied SMED principles to the maintenance of a polymer "classifier deck," which is used to control the size of plastic pellets from the extruders by passing them through two screens of different mesh size (see Figure 4.1). Oversized pellets and clumps of polymer are rejected by the first screen while properly sized pellets and small pellets or crystalline fines pass through. The second screen identifies properly sized pellets to advance in the process while fines and small pellets are rejected by falling through.

The screens require changing periodically due to mechanical wear. We were able to reduce the total lost time for this task from 6 hours to 6 minutes on the first attempt simply by aggregating the materials and tools in advance of stopping the unit and posting the instructions for the work as shown in Figure 4.2. Although the work of changing these screens is not complex and occurs about once each month, Exxon has five operating teams in order to provide coverage for the nonstop operation, so each team did the job relatively infrequently. As a result, team performance benefited

FIGURE 4.1
Polymer classifier deck in operation.

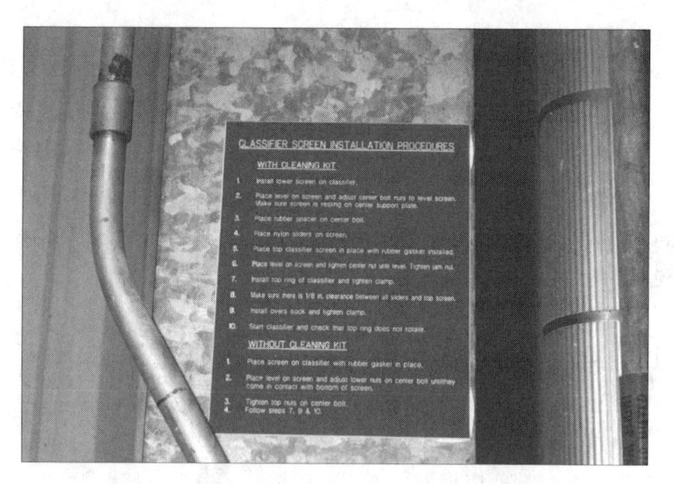

FIGURE 4.2
Instructions for maintenance of classifier deck.

greatly from simply staging the materials and reminding the team how the work is done.

Teamwork

A very valuable lesson gained from examining the actions of pit crews as compared to amateurs changing their own tires is that pit crews always use more than a single person. They appear to have one person for every possible place where work can be performed without one team member interfering with another. It is also obvious that the teams have a plan for working together whereby each person's individual work is always complementary to and coordinated with the other members' work. *No one is ever uncertain about who is doing what.*

As with preparation, using several people to do the work reflects the recognition that the *only* important attribute of a pit stop is to get the car back on the track. Given sufficient time, a single person could do all the work, but that would not be the best way to do it if you intend to win the race. In the process industries, we need to come to that same understanding. In most situations, a team of several people will return our equipment to service more effectively than a single person could.

Although the emphasis in this discussion is on the speed of returning the equipment to operation, do not be fooled into believing that using several people for a task that could be done by only one person is necessarily an inefficient use of labor.

Case Study: Reduce Downtime with Reduced Work Hours

Exxon once owned a business that extruded plastic filaments and then wove that filament into an inexpensive fabric used, among other things, to wrap bales of cotton. As part of that weaving operation, workers periodically needed to add several hundred new spools of the filament to the "creel" that fed the weaving loom. Each spool needed to be joined to the loom by tying the starting ends of the new spools to the trailing ends of the old spools.

Originally, this task was assigned to a single person, and it regularly required a full 8 hours to accomplish. When we assigned a small team of five people to the same task, they completed it in less than a half hour. In addition to returning the equipment to operation more than 7 hours faster, only half the total amount of labor was required.

There were many reasons for this difference. Among them, for a single person working alone, the work was solitary, boring, and physically demanding due to long hours of repetitive motion. For a small team, it was a nice way to spend a half hour. For a single person, the 8 hours included rest breaks and breaks for personal needs as well as for lunch. During those breaks, the equipment was out of production and no one was working to return it to service. For a small team, which was able to start and finish the work in a short time, none of those interruptions occurred.

Equipment

Equipment is the thing most people think of first when they consider the differences between the professionals of NASCAR and the poor souls standing beside their cars in the parking lot. Many professional racing cars do indeed have a value that far exceeds the cost of most personal cars, but the value is mostly associated with equipment that enhances performance on the track. Only a very small portion of that value is the result of the modifications that make it easier and faster to service the car. The cost of those modifications to enhance service is even smaller when considered in relation to the business value of keeping the race car on the track.

Exactly the same relationship is true in our plants. Although many SMED examples will be of the type shown in Figure 4.2, where there is little or no cost, there certainly will be times when you need to spend some money to get the SMED effect. More than half of the content of Shingo's book is detailed descriptions of simple modifications that you can make that will allow your equipment to be serviced or reconfigured quickly and easily. In every case, the changes that Shingo describes are of the same sort as the modifications that pit crews adopt to make race cars easier to service. They are always small relative to the value of the equipment and very small relative to the value of keeping the equipment in operation.

Once again, engineers and managers can never hope to work at the level of detail needed to modify all the equipment in a plant for this purpose, but when you adopt SMED as a formal practice authorized for autonomous action, you can teach its use to many people. Frontline teams can each modify their own equipment and their own work in modest ways according to the new practice, and you can expect to convert much of your plant rapidly to fast and easy mechanical work.

Case Study: Stimulate Small Event Improvements

At the Exxon Baytown chemical plant, when we first introduced the SMED technology, we had two well-respected technicians who promptly produced a sign that proclaimed their little shop on the plant floor to be the "SMED-quarters." They regularly modified equipment using this technology on their own initiative, which was within their existing authority. In addition to the operating improvements we received from their efforts, as recognized shop floor leaders, their buy-in led to even more widespread acceptance of the technology. It was obvious that they had been waiting for an opportunity to make a greater contribution and SMED gave them that opportunity.

Further, the synergy that is possible between the big event improvements of engineers and the small event improvements of frontline teams is apparent in this effort. When frontline teams make changes to their equipment, the engineers can adopt that practice and incorporate it into all new equipment in similar applications. It is also possible for engineers to adopt SMED as a big event project. In the plants of a major paper products manufacturer, nearly all equipment has the change-out parts needed for product transitions incorporated into modules of a standard size and configuration that are designed for identical positioning and assembly into the main line as a means to promote rapid exchange.

Once the line stops, the current modules are unfastened and pushed out of the way in one direction while a replacement module is simultaneously pushed in from the other side and assembled. No adjustments or resizing of the line or the components is required. In this way, product transitions can be completed very easily and very quickly. Each module is likely to be a compromise compared to a perfect design for that module, but, as a production system, it works well. This is an enormous improvement over the prior practice of designing each item to be perfect individually but imperfect as an element of a production system.

HOW TO USE THE SMED CONCEPT

Once you recognize the need, SMED is really simple and intuitive. I frequently receive feedback that people have previously, but informally, done something similar to SMED. The story usually goes like this: "Four years ago, we had an important job that needed to be done much faster than we ever thought possible. The techniques of SMED are exactly the same as the things we intuitively did." When I hear these stories, I am always surprised that so few people have recognized, as Shingo did, the opportunity to employ these practices formally and routinely through the actions of many people, rather than saving them for special occasions and the actions of a few specialists.

Case Study: Avoid Service Interruptions

A very interesting example of an early use of this concept is described in the book, *The Pennsylvania Railroad: A Pictorial History,* by Edwin P. Alexander. This book, published in 1947, described an event that occurred in 1901. The railroad wanted to replace a bridge over the Schuylkill River in

Philadelphia. To do that without interrupting this important transportation link, they erected temporary structures on either side of the existing bridge. On one of those temporary structures, they built a replacement bridge.

With all possible advance preparations in place, on Sunday, October 17, 1901, at 2:53 p.m., the last regular train crossed the old bridge. As soon as the last car of that train crossed the bridge, two teams of workers (one on each end) began to disassemble the connections linking the bridge over the river to the tracks on the land. Four minutes later, two special engines, which had been positioned in advance, went into operation. Using a series of pulleys, one engine pulled the original bridge out of position onto the second temporary structure while the other engine pulled the new bridge off the first and onto the bridge's permanent foundation. Three minutes later, the connections between the land tracks and the new bridge were in place. At 3:05 p.m.—12 minutes after the last train had crossed the old bridge—the first train crossed the new bridge. That was in 1901.

When I started incorporating this story into my examples of SMED practice, the chemical industry folks, who are regularly faced with extremely large mechanical work, began to talk about *SMEB* (or "single minute exchange of bridges"). That has been a useful addition to the original SMED concept for the process industries.

At Suncor, we recently had to exchange the entire convection section of a very large vacuum furnace that had eroded from the sand-laden oil pumped through our pipes. SMEB gave us the incentive to do that work much faster than we had ever considered possible. This was "found work" during an outage that had been planned for another purpose; nevertheless, by approaching the work in this way, we were able to undertake and complete this massive task with no impact on the original schedule.

THE FIVE KEY COMPONENTS OF SMED PRACTICE

Formal SMED practice has five separate components:

1. *Separation of activities:* To reduce the time that equipment is out of operation, separate the work of servicing, or reconfiguring the equipment into the "internal" tasks, which must be done while the equipment is out of production, and the "external" tasks, which can be done before the equipment stops or after it restarts. This separation allows us to limit the work during the downtime to only what *must be done* within that period.

Note: In the chemical industry, a much less detailed form of this sort of planning is used regularly in preparing for periodic "turnaround" events. The difference is that for turnaround planning, a few planners are assessing a 500,000-hour event plan in increments of 100 hours or more. In SMED practice, small teams plan their work in increments of minutes or even seconds.

2. *Modification of internal activities:* The period of the production outage can be reduced further by modifying the internal activities. One form of modification is to undertake a detailed analysis of large internal tasks to identify meaningful subelements that can be segregated from the rest of the internal work and done externally. A second form of internal modification is to remove waste from the work itself to make it faster to perform. The third method is to subdivide the work in a way that will allow more work to be done simultaneously.
3. *Modification of the work team:* Return the equipment to operation faster by utilizing an *appropriate* team to do the work rather than relying on a single individual or on the smallest possible team as is often the case in traditional maintenance practice. Utilizing the right team to perform many tasks simultaneously can often achieve an increase in speed that outweighs the increase in workforce. Indeed, an increase in speed often reduces the total work content of an outage. Appropriate modifications to the internal work increase the opportunity for more people to do more work in parallel.
4. *Modification of the equipment:* Once you implement components 1 through 3, move on to modifying the equipment. The great speed and value that can be obtained from making a few simple modifications to the equipment for the purpose of reconfiguring or servicing it more easily and quickly are truly amazing.
5. *Preparation for each event:* Preparing for SMED events incorporates the outcomes of the other four steps in the process. The goal is to execute the work precisely as planned, and therefore this is a very detailed step in the process and is necessarily very close to the action at the frontline of your business. I cannot say this too often: *Your frontline teams can succeed at this intense planning for all of your equipment and your professional staff cannot.*

We will now discuss each component of SMED in sufficient detail to allow you to implement it in your plant.

FIGURE 4.3
Replacement parts staged at the job.

Separation of Activities

The first step in SMED practice is to separate the activities associated with a production outage into "internal" work that must be done during the outage and "external" work that can be done before or after the outage. In many situations, once the SMED practice is well developed, it is possible to identify external work elements, such as aggregating materials and staging tools, that can be done in an essentially permanent manner so that they need not be redone at any time.

This concept of essentially permanent preparation also applies to preparing the detailed instructions for the task. Once prepared and properly documented as part of a SMED initiative, the work plan does not need to be redone until an improvement in the process occurs. It is also possible to stage consumable materials at the job site and refresh them after each event. (Figure 4.3 shows a kit of consumable parts permanently staged for servicing extruder die heaters and Figure 4.4 shows a technician with a small kit of tools appropriate to the work staged on the job.)

The objective of separating activities is to ensure that while the equipment is out of production, the only work done is the work that can only be done during that period. Any other tasks must be done *before* the equipment is taken out of operation or *after* the equipment has been returned to service. We routinely plan in this manner when we plan for big tasks as part of big events, such as turnarounds.

FIGURE 4.4
A technician with tools staged at the job.

Note: The *new* news is to practice this separation for small tasks as part of small events. This intense level of detailed planning requires engaging a new group of people to help.

Key idea: When people recognize that they previously used SMED concepts to improve an event, separating internal and external activities is a part of SMED they often describe.

Separation of activities begins by formally identifying in substantial detail all of the steps that go into the proposed work and then rigorously segregating them into work that must be done while the process is stopped and work that can be done while the process is in operation. This latter category includes work elements that can be prepared in advance of stopping the process as well as those that can be completed after the process returns to operation. Work that must be done while the process is stopped is labeled "internal" because it *must* be done within the process outage. Work that can be prepared in advance of the outage or completed after the equipment is restarted is labeled "external" because it *should not* be done within the outage.

FIGURE 4.5a
Spaghetti diagram before improvement.

FIGURE 4.5b
Spaghetti diagram after improvement.

Case Study: Conduct "Real Time" Reviews

If the work under study occurs so frequently that it is possible to observe it rather than analyze it in theory, that "live" review often has special value. A technique used in that situation is called a "spaghetti diagram." For a spaghetti analysis, a simple schematic of the work area is prepared and the drawing is annotated to track the physical movements of the technicians as the work is performed. As Figure 4.5a illustrates, the reason for the name is obvious. In that analysis, we found that for most of the time the equipment was out of service, the technician was moving around the work area, rather than working on the unit.

After improvement, most of which consisted of arranging the work area to make tools, parts, and information conveniently available, we eliminated in excess of 90% of the time it took to perform the work. The spaghetti diagram following the improvement is shown in Figure 4.5b.

Key idea: Technicians always respond to the unmistakable message of spaghetti diagrams. Following Suncor's first use of this tool, two technicians arrived for work the next day wearing pedometers.

Once all the work steps have been identified and each of the individual steps has been categorized as internal or external, the execution of the job is planned and scheduled so that only the internal work is done while the process is stopped. Because the goal is to reduce the period of the outage, the team must ensure that all possible external work is removed from the downtime.

This simple separation of activities often results in a huge benefit with no real application of technology or labor beyond planning and coordinating the work. Even in otherwise very good plants, I have seen countless incidents where the operating team would cease production, and only after the equipment was idle would they begin to do all the work of preparing it for mechanical service. As embarrassing as it is to admit, I have seen capacity constraining equipment in my own plants that was idle for long periods before people began to do the required mechanical work in earnest.

Case Study: Minimize Wait Time

Recently, I was driving through the Suncor plant with an area maintenance manager when one of the large ore preparation conveyors slowed and then stopped right in front of us. While we were on the radio trying to find out what had happened, a lubrication truck arrived and the oiler began to arrange his equipment in order to service the large roller bearings at the end of the conveyor. The unit was out of service for nearly an hour before the oiler finally was ready to begin the work.

In many plants, it is common to stop the equipment early to wait for the craft technicians. Apparently, the belief is that labor will be more efficient when the equipment is waiting for the technician than when the technician is waiting for the equipment. However, in capital-intensive industries, that appearance of efficiency is seriously misleading. As with the race car, the activity of principal value in a process plant, especially a capacity-constrained process plant, is operating the equipment to produce products.

Key idea: Efficiency in any activity that keeps rate-limiting equipment offline is probably counterproductive to the business.

Planning the work and organizing the work according to this internal/external classification scheme often results in very significant reductions in the total production outage. These improvements are generally available promptly and they are usually available without cost. Moreover, in formally planning routine work, even work done relatively infrequently has sustained value. With *documented plans* that are *permanently retained* and *regularly updated* based on new information or experience in the field, repetitive mechanical work can become routinely efficient and quite effective in rapidly returning the equipment to operation.

Planning in this way does not require a very formal process. Often, with some little formality, such as posting instructions on the wall, or a small amount of preparation, such as permanently staging required tools or parts on the job, the improvement that you can implement almost immediately and almost without cost is impressive and permanent.

Case Study: Analyze the Work to Be Performed

When we first assessed the pelletizing extruder to separate the activities, we found that much of the 15 hours of production outage was completely unproductive. One hour was lost in random delays after the extruder ceased production and before any work started. A full 2 hours were consumed by the mechanical crew's rest breaks, personal need breaks, traveling to and from the job in the field to have lunch in the shop, and during the transition as the shift changed. A further 2 hours were consumed by acquiring rigging equipment and hand tools, retrieving the change-out components from storage, and other things that could and should have been done external to the production outage. Analysis of the work itself suggested that a further 2 hours could be saved by organizing the work area in advance of starting the outage in a way that would reduce the amount of time people spent walking unproductively around the area.

After the initial analysis of the outage, we brought the craft crew to the area in advance of the shutdown, added sufficient people to the team to allow for staggered breaks and lunches, and held the first crew on the job until the relief crew arrived in order to allow for continuous work through the shift change. Starting work promptly and working continuously through the outage period saved 3 hours of production time. Moving the work of aggregating tools and materials outside the period of the outage saved another 2 hours. Arranging the work area to facilitate the work saved 2 more hours. All told, this simple first assessment of the production outage saved more than 40% of the original lost capacity.

Key idea: The effect of the work crew going to lunch during the outage is similar to that of the single worker restringing the fabric creels.

Once the time required to complete a task extends beyond the natural period that a person can work without a break, the period that the equipment is out of production often lengthens without regard to the needs of the work.

Modification of Rate-Limiting Internal Activities

Once you have gained the nearly free and nearly instantaneous benefit of separating internal work from external work, you next begin the process of improving the remaining internal work that limits the rate at which you can complete the service and return your plant to production. As you begin to study the nonroutine work of servicing or reconfiguring equipment as if it were routine production work, you will find many opportunities to improve or modify nearly all elements of the work to enhance efficiency in exactly the same way you improve the efficiency of other important production work.

Many of those opportunities will be to improve the efficiency of *external* activities. Those are good opportunities, but do not let them distract you from your most urgent work of reducing production losses by improving *internal* activities. The first improvements you make should focus on rate-limiting internal steps in ways that shorten the duration of the outage. If nothing else, demonstrating clear and consistent management focus on returning the equipment to operation continuously reminds the execution teams about their principal objective, which is always clear on the racetrack but often surprisingly unclear in manufacturing plants.

One easy modification of internal work often consists of subdividing a big internal task into several smaller tasks. Frequently, one or more of those smaller tasks can become external work. Alternatively, by subdividing an internal task into smaller elements, it is often possible for several internal steps, previously conducted sequentially, to be conducted simultaneously. When the detail of this analysis and improvement extends to activities measured in minutes or seconds, it becomes apparent why this is a job for the frontline teams who actually do the work.

Case Study: Question Assumptions

In chemical plants and refineries, our technicians routinely work on hot equipment. They already have the skills, special tools, appropriate safety practices, and personal protective equipment needed to work safely and effectively in that mode. As we conducted the original analysis and separation of tasks on the extruder, we believed that cooling the equipment and reheating it had to be internal steps. Because we knew that polymer could not stop inside a hot extruder, the first step in preparing for the work was to cease production and purge the machine. After that, without considering that we had the capability to work on equipment while it was hot, the shutdown team proceeded to cool the equipment.

Both cooling and reheating required a lot of clock time, during which the equipment was out of production. As one of our first steps toward modifying internal work, we adopted the practice of working on the extruder while it was hot. We had not previously used our ability to work on hot equipment because we were generally accustomed to "working hot" only when it was not possible or practical to take the equipment out of service. In this circumstance, once the extruder was isolated, the "in service" rationale for working hot disappeared in the minds of the craftspeople and planners. They simply assumed that it was practical to cool and reheat the extruder rather than work on it while it was hot.

Using an existing work capability in a new way changed the internal work and saved a further 4 hours from the total time required for the production outage. Again, there was no real cost to obtain this benefit. As shown in Figure 4.6, in the vented mode of operation, polymer routinely entered the vent assembly. Removing the vent after the infiltrating polymer had cooled was physically challenging because the polymer had solidified across the joint between the extruder and the vent. An added benefit to working the job hot was that cold polymer no longer interfered with the work.

FIGURE 4.6
Cold/hard polymer in the extruder vent.

Other areas where it is common practice to subdivide internal work in process plants are insulation, scaffolding, and heat tracing. Although each of these jobs is an element of other work, each can be moved externally to the outage in most situations. In fact, these activities are commonly done externally to the outage in planning for large jobs; however, on smaller jobs, the detailed planning is not sufficient to achieve that result and they frequently occur during the outage. In addition, you can often purchase some modest spare equipment that greatly changes the internal work. For example, a spare tube bundle for a heat exchanger allows you simply to swap bundles while the equipment is offline and clean the current bundle after the unit has returned to service.

Case Study: The Value of Small Changes

As shown in Figure 4.7, at Exxon we had several units with two heat exchangers in the same service. As we developed our SMED capability, we found that we could divert the full cooling flow to one exchanger, isolate and open the other, swap the bundle in service for our spare, and return the isolated exchanger to service sufficiently quickly that the process did not react to the fact that we were working on it. We then "externally" cleaned the bundle, which had been removed from the first exchanger, and again performed the fast swap process on the bundle in the second exchanger.

FIGURE 4.7
Two heat exchangers (top and bottom) in the same service.

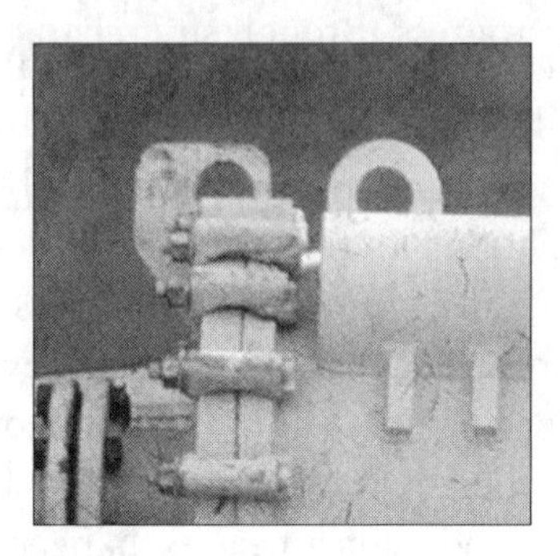

FIGURE 4.8
C-clamp flange.

What had previously been a multiple day outage—to stop production, open both exchangers, and clean both bundles—was transformed into one where there was no downtime at all.

If you look closely at Figure 4.8, you can see that this fast work was facilitated by a small improvement to the flange bolting. We converted from normal flanges and bolts to solid flanges with hinged C-clamps that served the same purpose quite well. This small change to the equipment greatly enhanced the speed of opening and closing the exchangers and contributed to our ability to work on the equipment so quickly that the process did not respond to the work.

Modification of the Work Team

As we have seen:

- Five people can restring a weaving creel in 1/10 of the time and with less than half the labor of a person working alone.
- Adding a few extra people in order to stagger breaks and lunch periods can shave 2 hours off converting an extruder by eliminating interruption.
- A pit crew can change four tires in a few seconds, whereas a person working alone needs a half hour to change a single tire.

When looking for ways to capture similar opportunities, you should focus on work that would benefit from *more* people doing some or all of the internal activities in parallel, resulting in a shorter production outage. For example, a small team might replace an individual or a large team might replace a small team. Alternatively, look for opportunities where a few extra people will allow you to shorten the downtime by keeping the work continuously in progress through shift changes, breaks, and lunch. By changing the assigned workforce in a way in which the internal work is conducted more quickly and more effectively, you can often return to operation much more quickly and often without a loss of efficiency.

Key idea: On work of extended duration, always structure the work team to ensure that the personal needs of the crew do not cause the work to stop or substantially slow down while the equipment is out of service. Personal breaks, lunches, and shift changes are all periods when this can easily happen if the work is not carefully managed. In chemical plants, where large-scale work often proceeds around the clock for extended periods, failure to manage the personal needs of the crew and transition periods between crews can result in an outage that is up to 30% longer than it needs to be to accomplish the intended work.

Case Study: Redeploy Staff during Outages

At Suncor, we routinely have crews who are continuously engaged in corrosion under insulation surveys, service of steam traps, heat tracing, and even large work such as cell-by-cell replacement of eroded piping in our furnaces. That work is important but not urgent. Therefore, we stop that work during outages and *redeploy* those people *to supplement the outage execution team,* which allows us to return the equipment to service much faster. Virtually *every plant has important but nonurgent work that can be stopped temporarily* to provide extra hands during an outage. Identifying that work *in advance* will let you make redeployments quickly when needed.

This practice had two additional benefits. First, by formalizing our understanding of the relationship of this work to other work, we began to make good progress toward completing our important but not urgent work, which previously had not received the attention that it deserved. Second, we found that, by redeploying our own people to outages rather than acquiring temporary workers, we had a much better and safer crew on the job and a much more stable workforce overall.

Modification of the Equipment

I placed this element of SMED practice fourth on the list for two reasons. First, because people like to modify the equipment as a way of creating change, I wanted to emphasize that often other methods provide a faster and cheaper way to start the process of improvement. Second, modifying the equipment often costs money and I am a firm believer that improvement ought to pay for itself. (If quality is not free, it should at least be self-financing.) Following this sequence for implementing SMED enables teams to accumulate the money—by time saved or, in the case of increased production, by profit earned—that they will spend in modifying the equipment. That practice of earning the money the team will spend is often surprisingly attractive at the frontline.

Key idea: At the beginning of any effort to increase frontline improvement activities, the improvement model that the teams possess is the big event model used by engineers and managers. Normally, that is the only improvement methodology that they have seen. By restricting or delaying the funds available for equipment modification, you can help them learn the small event improvement model, which is the essence of improvement at the frontline.

Shingo's book contains hundreds of examples of fast and easy equipment modifications. Prior to Shingo's death, I frequently spoke at seminars and conferences where he also spoke, and I learned that he enjoyed visiting a plant for a single day during which he would design modifications that were so fast and easy they could be conceived, built, installed, and used for a great improvement while he was still on site. I have retained that as a model for SMED practice. Frontline teams ought to be looking for modifications that are of a size that can be executed promptly. This is not the time for a capital project.

The work of reconfiguring or maintaining equipment is often a matter of removing one component and replacing it with another. Therefore, many of the best equipment modifications are changes that improve the way the various components are positioned for mounting and the way that the mechanical attachment is made. These changes are usually easy to conceive and fast to implement at relatively low cost.

Modify Equipment to Maximize Efficiency

The extruder example is a good illustration of this interchange of components. The equipment modifications made while implementing the extruder improvements are representative of the types of changes that you may want. There are many more examples in Shingo's book and elsewhere in which it is often possible to find an improvement that fits your needs so that you do not have to reinvent the wheel.

What follows is a description of only four of the several mechanical changes that we made.

Change One: Improve the Attachment Mechanism

I am not sure why people in industry use the phrase "overengineered" to describe things that have been designed to have far more capability or to be far more robust than needed, but that description has regularly cropped up in conversations for the past 40 years. In the case of the extruder, the overengineered equipment was the assembly for the interchangeable components that converted the extruder from vented to nonvented or plugged operation.

In the vented configuration, the extruder had an open flange that allowed connection of an exhaust that would draw chemical gases into a closed air-handling system to capture and control the gas rather than allow it to escape into the atmosphere upon exiting the extruder. Normally, this configuration was used when a chemical such as peroxide was added as a modifier to the plastic and we did not want the excess peroxide gas to escape.

When we first began to study this work, the vent was attached to the extruder using about 30 bolts, as shown in Figure 4.9. To attach the vent

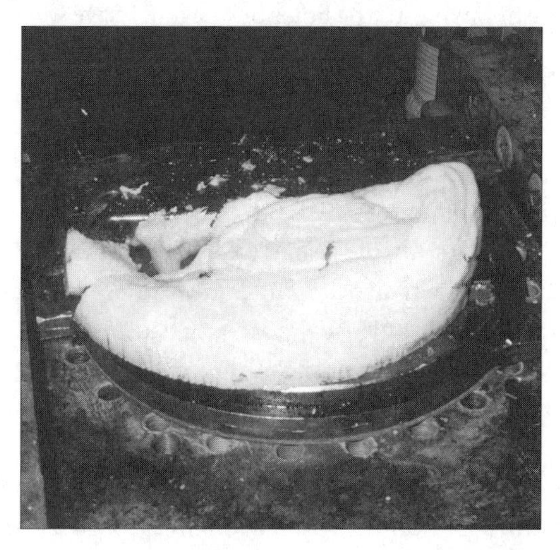

FIGURE 4.9
New configuration vent with two bolt holes sitting on an extruder body containing 30 attachment holes.

to the extruder, the change parts had to be positioned, the flange holes aligned, and a loose bolt inserted and then tightened in each of 30 holes. The slow and laborious process of disassembling the old piece and then positioning, aligning, and assembling the new piece required almost 2 hours. This did not include handling the change pieces (discussed later). We concluded that this assembly had been overengineered because while the extruder was in vented operation, there was no (zero) internal pressure on the flange; therefore, the assembly clearly did not require 30 bolts.

Note: This was an ad hoc engineering determination; we did not allow the maintenance frontline team to reduce the security of the flange autonomously.

As a result of this assessment, we made the first of a series of mechanical modifications to the extruder. We reduced the number of bolts used in the assembly of the vent from 30 to 2. Recognizing that the attachment as designed was far more rigorous than it needed to be allowed us to greatly reduce the effort required to assemble and disassemble it.

Note: We did not initially modify the machine; we simply stopped using all the bolts.

Change Two: Modify the Flange and Make It Self-Locating

Because we were not using all the holes on the flange, we were able to modify the flange itself. First, to eliminate confusion, we removed the flange segments with the now unused holes. This allowed us to modify the remaining flange segments so that the flange became self-locating. As shown in Figure 4.9, the remaining flange segments were squared and extended; thus, as the flange is set onto the extruder body and rotated, the flange segments engage with the body of the extruder and naturally position it properly for assembly. This extruder is a huge piece of equipment, and the vent is large, awkward, and heavy. In the past, it had been a time-consuming and physically difficult task to position the vent on the extruder and align the bolt holes for assembly; now, the vent positioned itself.

Changes Three and Four: Modify the Components and Use a Small Team

We made two other modest mechanical changes, both associated with the handling of the large, awkward components. In the original configuration, when a vent or plug was to be removed from the extruder, the attachment bolts were removed, the mounted piece was levered up with a pry

FIGURE 4.10
Original vent piece showing 30 attachment holes and rope sling.

bar, and a rope sling was attached through the bolt holes of the piece to be lifted (see Figure 4.10).

As we modified the equipment, in place of a rope sling, we permanently mounted rigging points on both the vent and the plug. (As difficult as it is to believe now, in one of the most technically advanced manufacturing plants in the world we actually kept a 70,000-pound-per-hour extruder out of production for more than a half hour while two men lifted the vent with pry bars and tied ropes to it!) In the modified configuration, attaching the vent to the hoist takes less than 1 minute while hoist hooks are quickly clipped to the permanent lifting points.

The second handling change was to discontinue use of a portable crane (shown in Figure 4.10) and to erect an overhead trolley rail and hang two chain hoists from the same trolley so that both pieces (the piece being removed and the piece being installed) could be handled at the same time. (Again, it is almost impossible to believe we had previously kept our extruder out of production while one component was lifted, lowered, and disconnected from the crane and the other was connected to the crane, lifted, and positioned!) We regularly lost 30,000 pounds of production because we did not have an installed hoist. (The folks on the Pennsylvania Railroad had solved that problem nearly 100 years before!) The cost to install the rigging points and to install the two hoists was trivial compared to the benefit of returning to production.

I learned this lesson again at Suncor. In order to achieve the rapid replacement on the convection section of a vacuum furnace (described previously), we deployed two 500-ton cranes—one to lift off the old section and one to place the new section. This step alone reduced the total elapsed time of the work by more than 10 hours because all the rigging and placement issues for both lifts were handled simultaneously rather than in sequence. The cranes were expensive, but not nearly as expensive as the production that would have been lost by extending the time our unit was out of operation.

The Result of Changes

In this revised configuration, the mechanical work required to convert the extruder is quite small, certainly much smaller than it had been. The task of converting the extruder from vented operation to plugged operation now proceeds as follows:

- Before the work begins, and with the extruder in full production, thermal barrier blankets are spread over the extruder to protect the technicians, who are also wearing appropriate personal protective equipment for hot work.
- All the components and tools necessary for the work are assembled and ready.
- The component currently attached to the extruder is hooked for rigging to one of the hoists.
- The replacement component is rigged and is hanging at working height adjacent to the extruder using the second hoist.
- The work begins by unbolting the two bolts holding the piece currently assembled to the extruder.
- The current component is lifted off the extruder by one technician using one hoist and the new component is immediately lowered into place by another technician using the second hoist. The new piece positions itself.
- The assembly is completed as each technician places and tightens one bolt.

At this point, the entire body of work has been completed. We simply change the material in the additive feeder and divert the small amount of material at the interface between the two products, and the conversion from one product to another has occurred with essentially no lost capacity and a very modest amount of work.

By working on the equipment very quickly—as we did on the two heat exchangers—our work on the extruder now has essentially no impact on production—an advantage that we have in the process industries that our manufacturing counterparts do not. Benefiting from the chemical characteristic of persistence, I have experienced hundreds of situations where significant production outages for routine mechanical work on process equipment were reduced to zero in this manner.

Note: This sort of improvement does not need to be labor intensive. Although one of the first steps in improving the conversion of this extruder was to increase the size of the maintenance team, by the time we were done, we were using less than 10% of the labor originally required for this work.

You will find more examples of modest but valuable equipment modifications in Chapter 9.

Preparing for Events and Sustaining the Improvements

The final SMED step is to prepare to use all the advantages that have been created. If the team that will do the work is the same team that has done the analysis and made the modifications to the work and the equipment, this preparation is generally very simple. If multiple teams are doing the same work, but the improvements have been planned and executed by only one team, or if team members have changed since the original planning, then the team that will do the work needs to rely on documentation of the new practices as it prepares for the execution of each event.

The critical management issue is that once an improvement has been achieved, it cannot be allowed to slip away due to the passage of time or a change in personnel. This is one reason that we cut the flange on the extruder. No team can slip back to making a 30-bolt assembly.

Key idea: There are two reasons why it is always good practice to document the results of improvements. First, in the process of making these improvements, you will have developed a good analysis of the work. Preserving that documentation will provide a basis for future improvements. Second, in many process plants, multiple teams are doing the same work and, although the work is repetitive, it may be infrequent, as was the case with the polymer classifier. Documenting the improvements enables every team to experience the same improvements without the need for every team to invent them. Similarly, undocumented improvements tend to slip away. *Clear documentation is a good reminder that an improvement exists that needs to be practiced.*

Outcome of Improvements

The outcome of the improvements to the extruder operation resulted in reducing the time that production stopped to allow this exchange of components from 15 hours to 0. The changes we made were all simple and even intuitive. Looking back, it seems odd that a fine business like Exxon had operated in the old manner for so long, but it is also easy to understand how the original 15-hour schedule occurred.

When the mechanical work required several hours and many loose pieces that might fall into the open body of the extruder, the desire to stop the operation is understandable. Once the decision was made to stop the extruder, it became necessary to purge the material. While the operation was stopped, it made sense to the technicians to work on it cold, which required further time. When the work, including cooling and reheating time, lasted for an extended period, it was relatively easy to add 2 hours to accommodate the personal needs of the crew for breaks and lunch and shift change. Without work analysis, it was easy to lose 2 more hours to simple inefficiency (spaghetti diagrams such as those shown in Figures 4.5a and b are always a surprise to the teams).

In my experience, every chemical plant has opportunities exactly like the ones that we have discussed. Big jobs like converting an extruder, cleaning heat exchangers, or replacing the convection section on a vacuum furnace all look like they should take a long time. Once everyone accepts that a job is likely to be a long one, it can grow much longer very quickly. As we experienced with our truck engine replacement, we recovered more than half the traditional lost time simply by paying attention to the clock as the job proceeded.

We certainly have found many of those opportunities here at Suncor. In some cases, the opportunities are valuable as the aggregate effect of small improvements in routine repetitive work, such as the improvements to the extruders and heat exchangers at Exxon and the trucks at Suncor. In other cases, the opportunity comes from a single big improvement in nonroutine work, such as the convection section replacement.

Sometimes these production outages occur to improve equipment flexibility when changing from one product to another. Sometimes the outage is to improve equipment availability during routine maintenance. Generally, even when production is rate limited, management does not attend or even closely observe routine activities such as reconfiguration of an extruder or cleaning a heat exchanger. It is much like paying bills or carrying out the trash. Managers know that it must be done, but it is not very interesting to watch and does not seem to offer a good opportunity to improve business performance. Adopting a tool such as SMED to enable the frontline teams to recognize the opportunity for you and to make the improvements autonomously is really the only way to deliver this benefit frequently to your business.

Industrial pit stops can be very interesting and extremely rewarding work for the people involved. The SMED practice can be the basis for establishing a great deal of employee engagement. People like using the new tool and they can be certain that as they do so they are doing the right thing in the right way to help improve the business that employs them.

5

Operational Planning to Improve Chemical Transitions

INTRODUCTION

Traditional lean practice focuses on the benefits to be derived from improving the mechanical flexibility of operations to transition more easily from one product to another. Due to the interaction between equipment and chemistry, some chemical plants experience an even greater need than mechanical plants for improved mechanical flexibility, while other process plants rarely experience it in a meaningful way. Despite differences among process manufacturers in their need for improved flexibility, the single minute exchange of dies (SMED) tool is appropriate to all chemical manufacturers because we can employ it to make a valuable contribution to maintenance and other mechanical work.

The technology described in this chapter for improving chemical flexibility is of value primarily to liquid manufacturers who experience problems during product transitions associated with reactive chemistry or chemical contamination. Here, too, process industries have a unique relationship to lean technologies because, when process operators do experience these problems with chemical inflexibility, the problems are often significant. Once again, the unique severity of our need represents an even greater opportunity for greater benefit than those achieved in mechanical manufacturing.

When we manage this situation well, we can make startling improvements. For example, one polyethylene plant of which I am aware had more than 200 products and a single world-scale reactor. In that situation, the company attempted to produce any one of its products at any time in immediate response to market demand. Although that might sound like a very lean operating practice, because of chemical inflexibility, the company was never able to produce more than 50% of its rated capacity. Its costs were among the highest in the industry and it was astoundingly unprofitable. The company had adopted the form of lean, but not the capability to make it work. The concepts in this chapter quite literally saved that business. Within 6 months of adopting these practices, it was profitable for the first time.

THE CAUSES OF CHEMICAL INFLEXIBILITY

It is often valuable to agree on the question before searching for the answer. Generally, chemical inflexibility results from two causes: chemical contamination and unintended chemical conversions. We will describe each so that as we discuss how to resolve these problems, it will be clear how the solution addresses the problem.

Chemical Contamination

Chemical contamination can occur in two different ways. The first source of chemical contamination occurs when a product of one type remains in the reaction vessel and becomes commingled with the next product in the vessel. The second source of chemical contamination occurs when a material, such as one or more additives or modifiers used in a reaction to differentiate the resulting product, remains in the equipment as the product changes, is present in irregular amounts, or commingles improperly within a single product. By whatever means the contaminating material arrives in the production vessel, the situation needs to be resolved before perfect production of new material can occur.

Unintended Conversions

Conversion of raw material into something other than the intended product is an experience shared by many reactive chemical operations. Although this conversion of material is truly unintended, generally, this outcome is not unexpected. It occurs during continuous production every time a product transition requires substantially altered process conditions.

Because it occurs during every transition, this loss has become a routine factor of production in much the same manner as mechanical transitions have historically been considered an unavoidable factor in discrete manufacturing.

Key idea: An unintended conversion occurs whenever the materials and the process conditions are not capable of producing the intended product, but are sufficient to produce a similar reaction that results in production of a different product.

Case Study: Unintended Product Caused by Changes in Reaction Conditions

In polymer production, it is common to produce polyethylene with propylene as a co-monomer. If one customer wants polyethylene with 3% co-monomer and another customer wants 5% co-monomer, both products will be produced in the same reactor during continuous flow. However, the reactor will naturally produce some polymer with 4% co-monomer as the reaction stabilizes to the new conditions. This is often commercially good material. It just is not what either customer wants. As a result, it often has far less value.

In terms of the lean values, chemical contamination during product transitions results in additional cost, the waste of material, and the loss of productive capacity. Production of unintended materials during a product transition also increases cost, wastes material, and reduces productive capacity. If the process manufacturer attempts to amortize these losses through long production runs following each transition, the added waste of inventory also occurs. Although these losses arise in a way that is unique to the liquid industries, their impact is identical to those of losses of a mechanical transition in discrete manufacturing.

FIXED SEQUENCE VARIABLE VOLUME PRODUCTION

The management practice that minimizes the impact arising from these problems that are unique to the process industries is fixed sequence variable volume (FSVV) production. In the simplest terms, FSVV organizes a fixed production cycle that sequences one product after another in a way that minimizes the number of transitions that have high cost and maximizes the number of transitions that have low cost.

Managing production in an optimum fixed sequence ensures that each individual transition is conducted routinely with minimum loss. That practice simultaneously ensures that the total losses throughout the full production cycle will be far less than in traditional operating modes, which randomly transition among the several products in response to an inventory management algorithm. Inventory-driven methods of production management barely consider the impact of product transition losses beyond increasing inventory as a way to amortize the loss. In fixed sequence operation, the total transition losses are purposely managed to be lower, and therefore the total cost of accommodating those losses with inventory is also lower.

Key idea: Managing a fixed production sequence to avoid transition losses as opposed to managing a random sequence of production to replenish inventory produces a perfect alignment with lean values and clearly focuses the entire organization on correcting the underlying problems rather than accommodating the symptoms of those problems.

The Concept: A Comprehensive Approach to the Production Cycle

At first glance, the FSVV concept of operating according to a fixed sequence of production sounds as if it is going to make your operations less flexible than they are today. It is certainly true that in order to practice a fixed sequence of production consisting exclusively of optimum transitions among several different products, it is necessary for each individual product to wait its turn. For plants accustomed to producing a product whenever they want, this requirement must seem like a step backward. Critical to comprehending the value of FSVV is to consider the entirety of your production cycle, rather than individual products within the cycle.

In an operation where all products are produced so that each product experiences the best possible transition, it is clear that the total amount of transition losses (including lost capacity) will be the smallest amount possible. The accommodating production of excess product placed into inventory will also be the smallest amount possible. Because during each cycle the capacity lost during transition is reduced and capacity that is consumed to make inventory is reduced, you effectively have more real-time capacity available within each cycle to produce the products currently required by the market.

With that increase in available current capacity, your entire portfolio of all products can be produced much faster and each product can be produced much more frequently. By purposefully optimizing transitions, you will reduce the wastes associated with inflexible transitions and you will immediately begin to notice an effective improvement in your ability to deliver all of your products as well as a nice reduction in the cost of production. Because you are naturally able to produce products more frequently, your practical ability to produce different products when you need them improves, rather than worsens.

What We Can Learn from the New York Subway System

As with SMED, it is useful to introduce the concept of FSVV with a simple analogy that will give you an idea of the big picture of this practice that you can refer to as you learn the practical details. In this case, we can use the New York subway system.

Compared with the convenience of stepping from your home or office at any moment into a private car that will take you directly to your destination, the subway seems like a pretty slow and inflexible mode of transport for each individual trip. You need to wait on the platform for a train that comes on a specific schedule. However, it is clear that the New York City subway routinely delivers all the people to all their destinations much more quickly and flexibly and at a much lower cost than would be achievable if everyone tried to move around the city in private cars. The practical benefit of the subway on all transportation in New York has been demonstrated repeatedly whenever subway operation was disrupted for some reason.

Although the subway appears to be a slower, less flexible form of transportation that has been imposed upon individuals for the greater good, it is in fact voluntary. Each individual standing on the platform awaiting his or her turn has made the choice that his or her trip, or at least the sum of all trips, will be faster, less expensive, or both because of the subway than the same bundle of trips would be using a private car.

Note: This is the first effect that we want to achieve with FSVV. For routine operations, we want to schedule the entire sequence of production in a predescribed way that reduces the losses that would otherwise result from randomly sequencing products as if manufacturing each product is a one-time event unrelated to other production.

In addition, because the subway manages the bulk of routine personal transportation, it creates a real opportunity for individuals with an instantaneous specific need for greater speed and flexibility to achieve that result by effectively using a private car. Without the subway to keep the roads clear, private cars would be useless in New York City even when they were most needed. This effect has also been well demonstrated during times of subway system disruption.

Note: This is the second effect that we want to achieve from FSVV. When the production or market situation truly demands a nonroutine response, the capacity and flexibility created from FSVV operation allows us to respond well.

FSVV production does exactly the same thing for your plant that the subway does for New York City. It organizes production into a fixed and scheduled system that routinely manages the bulk of your operations in a way that makes production of the entire portfolio of products faster, more flexible, and less expensive. Individual products must indeed periodically wait their turn, but all products are produced faster, more frequently, and at a lower cost. In addition, the existence of an FSVV system that organizes and speeds the bulk of your production helps you to manage effectively the necessary exceptions to this routine as they occur.

Key idea: The rigorous structure of the FSVV system may seem pretty far from the lean ideal of producing any product in any quantity at any time. It is important to remember that continuously moving toward the ideal is beneficial, but jumping immediately to the theoretical end state often causes more problems than it solves. FSVV is entirely consistent with the lean ideal. The new flexibility certainly moves us a great distance in the right direction and enables an effective practice of continuous improvement to complete the journey.

THE FOUR COMPONENTS OF FSVV PRACTICE

FSVV consists of four elements:

1. *Fixed sequence:* The enabling concept of FSVV is that some transitions necessarily cause great loss and others cause small loss. By establishing a fixed sequence of production, you maximize the number of easy transitions and minimize the number of hard transitions throughout your entire portfolio of production. In this way, the cost of each individual transition will be the least it can be, with the result that the total cost of all transitions throughout your entire portfolio will also be the least it can be.
2. *Inventory policy:* The FSVV inventory policy supports the fixed production sequence by providing enough material to meet customer demands throughout the cycle in a way that will allow the production cycle to proceed without interruption. By structuring the inventory in accordance with the fixed sequence, the amount of material produced for inventory continuously responds to improvements in your operation in a way that continuously makes the cycle faster.
3. *Variable volume scheduling:* Products are produced in different quantities—variable volumes—each time that they are produced in a fixed sequence. In traditional production scheduling, products are produced in a fixed volume on a variable sequence. The difference that results from this reversal of priority is a great increase in both the stability of operation and the effective flexibility of response to market demand from production rather than from inventory.
4. *Continuous improvement:* FSVV changes the playing field for continuous improvement of chemical operations. For a plant with 200 products and a single reactor, random sequencing of production implies the possibility of 39,800 different transitions. In that operating mode, management will be lucky beyond its dreams if the reactor teams regularly make the transitions correctly and can have no real hope that the teams will ever make the transitions better. In fixed sequence mode, a plant with one reactor and 200 products needs to manage only 200 transitions. As a result, operating teams should certainly be able to make those transitions properly and a very real

possibility exists for focused improvement efforts that will continuously make the transitions better.

Each of these elements of FSVV is described in detail next.

Typical Operating Problems

In this chapter, we will use examples based on a plant in continuous operation because this is the most complex situation. Batch manufacturers can easily see how to adapt this technology to their own use. We will begin with a brief introduction to the issues that will be used to illustrate the operation and benefit of FSVV.

Changes in Process Conditions

Many of the product transitions in reactive chemical manufacturing require changing the conditions of the reaction—for example, adjusting the temperature and pressure and residence time of the reaction. The transition losses could be relatively small if the changes in process conditions are small or severe if the magnitude of the changes is large. The most common source of loss during changes in process conditions comes from material that experiences an unintended conversion before the reaction stabilizes to the new conditions. If the change in conditions is small, then the time required to stabilize at the new conditions is also small and only a small amount of unintended material is produced. As the magnitude of the change in conditions becomes larger, the amount of unintended material produced becomes greater.

Additives and Modifiers

Other transitions require incorporating a variety of chemical additives or modifiers into the product or changing the amount of an incorporated material. The magnitude of the losses resulting from these changes is modest if there is only a single material and the change in the amount is small. The losses grow as the change in quantity of that single material increases. The losses are small if the change is between different incorporated materials that are mutually compatible. Conversely, the losses are large when a combination of materials is incompatible.

Changes in Reactive Chemicals

Changes in co-monomers and catalysts are representative of circumstances where product-to-product transitions require a fundamental change in chemistry. In this situation, the transition losses are modest if the materials are sufficiently compatible, but the loss could become severe if the materials are sufficiently incompatible. In some cases, the extent of incompatibility between different catalyst systems was so severe that no tolerance for any degree of possible contamination was allowed.

Key idea: It is clear that the cause of these losses is knowable and predictable. Although arising from chemical sources, they are identical in character to the losses arising from mechanical transitions. More importantly, these losses are just as manageable. The losses arise as a proportional response to the severity or ease of the individual changes required to produce the several products. Therefore, transition losses can be controlled by arranging the sequence of production in a way that minimizes the severity of each individual transition. Whether you have five products or five hundred products, it is possible to identify a sequence of production that routinely avoids or eliminates the most severe transitions.

The Fixed Sequence

The essential attribute of FSVV operation is the fixed production sequence. Similarly to SMED, FSVV is amazingly simple in concept and in practice once you understand the causes of poor chemical transitions. In relative terms, some of your product-to-product transitions are always easier and some are always harder. Some transitions are so difficult that they require extreme effort or care, which makes the successful practice of these transitions vitally important in the same way that critical equipment is important to your mechanical reliability.

Other transitions occur far too frequently because of poor inventory policies or unpredictable demand. These too-frequent transitions may be difficult or easy, but the attribute that makes them disruptive to your plant operation is that they occur more frequently than they should. This makes them the product equivalent of equipment bad actors. Just as we practice maintenance in a way that gives special emphasis to the care of critical equipment and the avoidance of bad actors, FSVV allows us to give special emphasis to improving product transitions.

Establishing a Fixed Sequence

A fixed sequence production cycle organizes the several products that share common equipment so that they routinely follow one after another so that each transition between two products requires the smallest possible change. In some cases, special insight is needed to achieve this result. For example, it might be best for a particular product to appear more than once in each cycle if the characteristics of that product enable many other transitions to be easier than they otherwise would be. Conversely, you might have to accept a suboptimum transition to one product if that transition makes many other transitions better than they otherwise would be.

Once established, the sequence of production proceeds in a series of optimum or nearly optimum transitions from product to product and returns again to produce the first product. In this way, the cycle of all products experiences the absolute minimum loss achievable with your portfolio. This form of organizing a production to run through the product portfolio in a repeating cycle is like a circle and is often referred to as the "production wheel."

Case Study: Constructing a Production Wheel

The easiest way to describe how a production wheel is created is to share the example of how we actually did it in Exxon's polypropylene plant in Baytown. The most important parameter that affected our transition losses was the catalyst used to initiate the reaction. At the time, we used three different catalysts. Although we also had various co-monomers and other reactive materials, for simplicity, catalyst changes will be used to represent the whole category of changes in the reactive chemicals.

The second parameter that had an impact on our transition losses was the melt flow index (MFI), which measures the rate at which a polyolefin will flow under certain conditions. For primary plastic manufacturers, MFI is used as an indirect measure of the polymer's molecular weight. Although operating conditions had an impact on many different parameters, for simplicity, MFI will be used to represent the loss category of change in process conditions.

The third parameter that affected transition losses was the incorporation of either a chemical modifier or an additive into the product. For these purposes, a chemical modifier is a secondary reactive material that is chemically incorporated after the polymer is formed in order to create a new chemical effect. An additive is a nonreactive material mechanically incorporated into the polymer after it is formed in order to serve a function that is largely independent of the basic structure of the polymer. Although various additives and modifiers are incorporated in different ways, for simplicity we will discuss the loss category of additives and modifiers as if they are all mechanically incorporated.

Different relationships among these three parameters influence the magnitude of transition losses. At the low end of cost, a small change in the conditions of reaction between two different melt flow indexes with all other parameters remaining constant normally results in the production of a small amount of unintended material. Also near the low end of transition cost is material that contains a single additive that is briefly present in a modestly irregular amount as a result of a transition between two products that have very similar specifications for that additive.

Near the middle of the spectrum of transition loss is material containing a mixture of two different additives. Also near the middle of the spectrum of transition cost is material that has significantly irregular MFI as the result of a large change in the conditions of the reaction.

At the high end of the spectrum of transition costs, some additives or modifiers are very incompatible with one another. Also at the high end of the range of transition losses are complex transitions where multiple process parameters are changing at the same time—for example, changing an additive at the same time as the process conditions change and as the co-monomer is modified.

At the extreme high end of transition costs are the transitions from one catalyst system to another. This is always a long and slow process. In some cases, the extent of incompatibility is so severe that the reaction is quenched and the reactor is opened and physically cleaned as if this were a batch process.

Of course, we faced other considerations, but those technical details are unnecessary for our purpose here. Using only these three representative characteristics (catalysts, additives, and MFI) to describe the production wheel for polypropylene is a simplification that allows us to focus on the concepts of FSVV rather than on the technology of polypropylene production. You will have to assemble a similar list of the operating parameters that are important to your business.

Faced with these three parameters, we took the following steps:

Step 1: Catalyst systems represented nearly all of the extremely high cost transitions that we knew we wanted to avoid or minimize promptly. Therefore, the first constraint we considered in establishing our fixed sequence of production was to establish a sequence that would restrict catalyst changes to an absolute minimum. Because we had three catalyst systems, we decided to allow each catalyst to be on the reactor only a single time during the period of a full cycle through all of our products.

Step 2: We first arranged the sequence of additives and modifiers to accommodate the way that they related to one another. When two additives could easily flow into one another with little or no consequence, we sequenced them together. When two additives were mutually incompatible, we often placed a product with no additive into the sequence to ensure clean separation between the two additives. We were more willing to accept irregular amounts of a single additive than we were to accept a mix of two incompatible additives. Within each additive family, we generally started with products that required a low amount of that additive and arranged a sequence that built to products that required high amounts of additive. We continued by decreasing the amount of the additive to move back to products that needed a low amount. We would then transition to a product with no additive or with a low amount of another additive. Because this sequence allowed us to transition continuously from a low amount of one additive to a low amount of another, we normally achieved these transitions with very little loss. In that manner, we avoided large changes in the amount of any single additive and we were generally able to avoid mixing more than one additive.

Step 3: Meeting customer specifications for the MFI parameter generally required that we change the operating conditions, such as the temperature, pressure, or residence time of the reaction. We wanted a sequence that would minimize large changes in reactor conditions. Therefore, we conducted reactor campaigns that progressed incrementally from high MFI to low MFI and back. In making additive changes, we started each family at a low value and cycled to high values and back to low values within a single additive family. In changing reactor conditions, we initiated a new family of products at the low end of the MFI range or at the high end and cycled through that family of products only until we reached the opposite end of the MFI range. At that point, we transitioned to a new family of products.

There was no value in cycling MFI from a low value to a high value and back again to a low value within a single family of products because we were not adding or withdrawing a physical material that needed to be cleared from the reactor, as was the case with additives. By cycling MFI between individual products at the extremes of reactor conditions, we minimized the number of times that we changed conditions, which reduced the extent of the losses derived from that effect.

We visualized this practice as producing our product wheel in blocked "campaigns" of catalysts, overlaid with "waves" of additives that further overlaid "ranges" of MFI.

As you might imagine in a plant with several hundred products, three catalyst systems, many additives and modifiers, and very many MFI values, achieving a fixed sequence was not the work of an afternoon and we did not get it exactly right the first time. In concept, though, the bones of what we ended up with and the way that we achieved it are in the preceding description.

- Within each product cycle, we had three big campaigns, one for each catalyst.
- Within the campaign of each catalyst, we cycled through waves of the various additives and modifiers according to mutual compatibility and used product with highly compatible additives or product with no additives at all to provide some relief within the cycle in order to avoid bad combinations.
- Within each additive wave, we produced products with a range of MFI values from high to low or from low to high by changing reactor conditions.

Initially, it took time to establish the optimum fixed sequence and the planners and engineers have constantly fiddled with it to make it still better ever since. However, the first attempt produced tremendous value.

- We were severely capacity constrained and immediately gained capacity.
- We had suffered from high costs; now our costs were among the best.
- We had previously damaged our reputation as a quality supplier and damaged our market pricing by selling a large amount of off-specification product with irregular amounts of additive or irregular molecular weights. That off-spec product attracted very low prices and its presence in the market had a carryover effect in holding down the price for all our products. As we proceeded with this effort, customers began asking for more of this "commodity" material than we had to offer and that off-spec product as well as our other products increased in value accordingly.

As further testimony to the prompt benefit of this practice, in the polyethylene plant with 200 products and a single reactor described earlier, they were attempting to produce any product, at any time, and in any quantity in direct response to customer demand. The resulting transition losses cost them dearly. Halfway through the year, they adopted the FSVV sequencing technology—without the other FSVV changes described here—and they reduced transition losses by 90%, increased reactor utilization from less than 50% to more than 85%, and became profitable for the first time ever.

FSVV Inventory Policy

The second element of FSVV is an inventory policy, the primary purpose of which is to ensure that the fixed sequence operates effectively and with a minimum of exceptions. To do this, the FSVV inventory policy intends to provide enough material to allow the sequence to operate without interruption caused by a market-mandated break in the schedule. In this instance, the inventory does not accommodate and hide the existence of the problem; rather, it is visibly and directly linked to unwinding the problem.

The second purpose of the FSVV inventory policy is to reduce transition losses further by periodically removing some products from the production sequence when that removal can be accomplished easily and with very little accommodating inventory. The inventory policy allows the operating team to take low-volume products completely out of some production cycles. The policy also allows the operating team to withdraw especially difficult products. This is similar to the maintenance concepts of critical equipment and bad actors that receive special attention. In this case, products that are too troublesome or too frequent are intentionally removed from the sequence. When there is no change of product, there is no transition loss.

The third purpose of the inventory policy is to facilitate continuous response to improvements in the cycle time in a way that reduces inventory as a means to lock in the improvements. Because the FSVV inventory policy produces products in amounts that are directly related to the duration of the production cycle, as the cycle improves, the inventory naturally diminishes in response. The improvements are sustained because no inventory is available to allow the cycle to revert to old practices.

We will look now at more traditional inventory policies so that we can compare them to FSVV and discuss the relative benefits of the FSVV inventory policy.

Days of Demand in Inventory

The most common form of product inventory policy is based on the calendar. For example, many businesses adopt a policy to hold in inventory

30 days of expected demand for each product. Because this policy is measured in calendar "days of demand in inventory," it is often called a DDI policy (e.g., a 30-DDI policy implies that there are 30 days of anticipated demand in inventory). Because a standard DDI policy applies equally to all products, it results in a lot of inventory for high-volume products and a little bit of inventory for low-volume products.

In practice, using this production management system, products are randomly produced whenever they are required in order to replenish inventory. In this scheme of production scheduling, all products are typically produced in a predetermined minimum amount. This is because in randomly sequenced production, it is impractical to calculate a production volume that will properly amortize the actual costs of all the possible transitions as they occur.

ABC Inventories

An alternative to the standard DDI inventory policy is a policy that modifies DDI to account for the importance or volume of a product. This is often done using a classic ABC approach to establish the categories of importance. In this scheme, "A" category products are important products with high volume/high value or products that are sold to an important customer. In other words, these are products where there are significant business consequences to running out of material; therefore, extra protection is held in inventory to ensure that this never happens.

At the other end of the scale, "C" category products are those with low volume/low value or sold to customers with no special business significance. There are few consequences to running out of these products, so there is less need to hold material in inventory to protect against that occurrence.

"B" category products are those that fall between the two extremes; they are held in inventory in accordance with some simple form of a standard DDI policy. The ABC inventory policy distorts the amount of material in inventory to protect the high-value products and customers. Rather than holding an equal number of calendar days of inventory for each product, there is a relatively larger amount of high-volume/high-value products and a relatively lower amount of low-volume/low-value products. Because of the impact of large inventories for the high-volume/high-value products, this practice normally results in generally higher overall inventories—although some planners skimp on the B category to bring the total back in line. Again, in this form of production planning, each product is randomly produced to replenish whatever inventory target has been assigned. Again, the replenishment amount is determined exclusively by the minimum production quantity assigned to that product because random production sequencing will not allow for precise determination of an optimum amount.

Key idea: It is important to recognize that in both of the most popular inventory management practices, the inventory determines the production schedule and transition losses are an unmanaged consequence of whatever schedule results.

FSVV Inventory Policy

The FSVV inventory policy differs from the DDI and the ABC forms of inventory management in two major ways. First, rather than plan inventory based on a fixed number of calendar days and produce an amount of product determined by a fixed minimum run length, the FSVV inventory policy plans both inventory and production quantity based on the time required to cycle through the fixed sequence of production. When a product is scheduled onto the reactor, FSVV produces enough material to last until the next time that the product is expected to be on the reactor.

Second, FSVV inventory policy distorts the inventory in the opposite direction from the standard ABC approach. With FSVV, inventory consists of relatively more of the lower volume products and relatively less of the higher volume products. The reason for this disproportional inventory approach is that it enables the production planners to drop low-volume products out of the production cycle at a relatively low cost for inventory. The result of removing low-volume products from the production cycle is a faster progression through the cycle so that high-volume/high-value products are produced more frequently. We protect our important products and customers not by holding a lot of material in inventory, but rather by greatly increasing our opportunity to produce those products efficiently. As a result, we provide greater protection for our important products and customers and we greatly reduce the amount of materials in inventory.

Using the ABC inventory policy definitions described earlier, a typical application of the FSVV inventory policy is as follows:

- For category A products, during each cycle, we will produce the inventory required to last through one complete production cycle, plus a little for safety.
- For category B products, during each cycle, we will produce the inventory required to last through one complete cycle, and there is no need to allow extra for safety.
- For category C products, each time that the product is produced, we will produce the amount required during two or even three cycles through the fixed sequence.

According to this FSVV scheme, for each class of product, you will produce and hold in inventory only the minimum amount of product required to reach the next production run for that product. Category A and B products are expected to run each cycle. Category C products run every second or third cycle. With a relatively small amount of inventory, you can drop a lot of products (and a lot of product transitions) out of each production cycle.

Key idea: Following a traditional inventory policy, products are produced in a random sequence triggered by the need to replenish inventory and transition losses are an unmanaged consequence. Following the FSVV inventory policy, every product continuously falls within its proper place in the sequence. In this case, the fixed sequence determines the transition losses and the inventory is a carefully managed outcome.

Establishing a fixed sequence for production in this way gives your plant the fastest possible cycle that will produce all of your products with the minimum of transition losses. Adding this inventory policy to the benefits of a fixed sequence makes your production cycle even faster and reduces still more of the transition losses. Because the resulting inventory is directly linked to the production cycle, this ongoing reduction in the duration of the production cycle immediately translates into a continuous reduction in inventory and a consequential further reduction in the duration of the cycle.

There are other benefits. The quantity and value of your inventory are often greatly reduced because you hold far less of your high-volume/high-value product in inventory. Despite that reduction in inventory, with just a little safety stock to protect against the variation in customer demand that occurs during the period of a production cycle, your ability to serve the market is well protected because you produce the important products far more often.

Note: Lean theory allows you to protect against problems that are out of your control and therefore impossible for you to correct. A little safety stock to accommodate unpredictable variation in customer demand for high volume/high value products that occurs during the production cycle is fully acceptable.

Lower volume products normally experience the highest variation in demand, but in this system, that result is fine. According to FSVV inventory policy, you normally have the highest amount of inventory for the lowest volume products. Further, if demand for a low-volume product is unexpectedly high, you have frequent and naturally occurring opportunities to add that product back into the production schedule each time in its proper position in the sequence. If demand is unnaturally low, you simply leave the product out of the sequence for another cycle.

Wheels within Wheels

A further benefit of the FSVV inventory policy is that it enables what has come to be known as "wheels within wheels." The original fixed sequence was described as the production wheel, so you can imagine what this is. Some additives or modifiers, chemicals, or catalysts are particularly nasty and cause serious problems with incompatibility and transition losses whenever they appear. Some additives, catalysts, or other combinations are developmental or otherwise new to the market or for other reasons have unusually low or unpredictably episodic volumes. Following an FSVV inventory policy allows you to place all such products into a special "D" category. (Some folks enjoy this designation and begin to refer to these as the "dog" products.)

The benefit created by assigning an entire family of products into a special production category is that the entire family can be taken out of the production sequence for one, two, or even three cycles according to your choice. Whenever one of the products is required, you start the wheel within the wheel and produce them all and then take them all out of the sequence for one or more cycles. The entire family of products occupies a single place in your production sequence that normally would be assigned to an individual product. Your fixed sequence ensures that you produce that family in an appropriate place in the production cycle and your inventory policy helps you—to the greatest extent possible—avoid the high transition cost of producing them frequently.

Variable Volume Scheduling

Until it is practical for a plant effectively to produce any product, at any time, and in any quantity, the ideal amount of a product that must be produced each time it enters production is the precise amount needed to amortize the specific transition losses required to produce it in a way that results in the cost or capacity your business constraints require. Unfortunately, in plants that allow random sequencing of production, the number of potential transitions is so great that it is not possible to practice this ideal state because doing so would require recomputing the transition losses for each production run. Instead, most plants simply assign a minimum run length to each product or each family of products.

Key idea: I regularly disparage the idea of "amortizing" transition losses as a production policy. However, so long as you experience significant transition losses, it is necessary to calculate the production quantities appropriate to your current losses that will result in productive capacity and product costs sufficient to keep you in business until you can successfully improve your transitions.

Following a traditional production planning process requires producing a predetermined fixed quantity of all products whenever inventory policy or market demand requires. In that mode of operation, the amount of material that enters inventory is always unrelated to future market demand. As a result, the time required to remove the material from inventory is uncertain and unrelated to future production planning. The FSVV production sequence and the FSVV inventory policy enable you to produce a variable amount of product during each production run in a way that makes inventory consumption and future production planning as predictable as they can be.

Case Study: The Inventory Impact of Fixed Volume Production

In 1987, prior to introducing FSVV, Exxon's polypropylene plant did not attempt to calculate the actual cost of each transition as it occurred. It simply established a "minimum run length" of 800,000 pounds of product for any product that was produced; in round numbers, this is the amount of polypropylene that will fill four railroad hopper cars. Although some high-volume products clearly demanded and received more production each time they ran, the majority were produced to this fixed quantity.

Key idea: Although scheduling according to a minimum run length often does cause low-volume products to drop out of production cycles in a manner that appears similar to the FSVV effect, in traditional run planning that effect is unmanaged. Products return to the production cycle in a random way and therefore are not produced in an optimum sequence or an optimum quantity. Furthermore, because transition losses are high and the related minimum runs are long, the inventory and related costs for low-volume products are always far higher than with the FSVV plan.

An FSVV operation has two critical differences from a minimum run length policy. First, by operating according to the fixed sequence, total transition losses are greatly reduced in a way that naturally reduces any prior minimum quantities based on prior transition losses. Of great importance, the specific transition loss for many products produced in the proper FSVV sequence becomes so small that those products do not properly have a required minimum run. They can be successfully produced in whatever quantity you desire, which is a real breakthrough on your path to lean operation.

Second, by establishing and following a fixed sequence of production rather than a random sequence, it is relatively easy to determine the exact production loss for each transition and therefore relatively easy to determine the precise amount of production required to amortize those losses in a way that will achieve your required productivity and costs. Products with transition losses that truly demand a minimum run in excess of the natural duration of the production cycle can be assigned to product category C or D in a way that properly produces enough material to take them out of the cycle and accommodates their abnormally high transition difficulties. It is possible to provide special attention to critical products and disruptively frequent products in a way never possible in traditional operating modes.

Key idea: Once you are able to stop producing arbitrary minimum quantities or quantities that are artificially inflated by very large transition losses, you can begin to approach the production of very precise amounts of any product. Often that amount is different for each product and even different for the same product from time to time. The critical issue is that the "big event," where managers and engineers create and enact the fixed sequence, causes an immediate improvement. That further provides a new opportunity for many future "small events," where frontline teams can continuously and autonomously create further improvement by optimizing transitions and the resulting inventory.

The opportunity to produce variable volumes comes into play in FSVV operations in three principal ways. The first is due to the operation of the fixed cycle itself. The FSVV inventory policy requires production volumes in increments, which, for each category of product, will produce material sufficient to meet expected customer demand during the period of the fixed sequence. Two characteristics of the system have an impact on this quantity.

First, category C and category D products will be continuously moving into and out of the cycle, and, as transition losses constantly improve, the cycle continuously becomes shorter. During routine operation of the production wheel, each time a product is scheduled onto the reactor, the planner needs to look at the expected duration of future cycles and the expected demand that will occur during those cycles and schedule a production quantity appropriate to the upcoming cycles as they will actually occur. This is a very different task with a very different effect than simply scheduling a random series of minimum quantities.

The second opportunity to vary production volume occurs when the presence of a product in the production cycle improves the transition losses for other products. A C category product in production now normally may be expected to drop out of the sequence for the next production cycle. However, when the planner examines the details of the transition sequence for upcoming cycles, he or she may see that this particular product plays an important role in limiting the transition losses that would otherwise occur.

For example, the direct transition from product X in the sequence to product Z in the sequence may have a very high cost during cycles when product Y is not produced; that same transition from X to Z may have a relatively low cost if product Y is produced between them. In such a case, recognizing that the objective of this technology is to reduce transition losses, every time that both products X and Z are scheduled for production in the same cycle, product Y will be scheduled to run between them. In this situation, product Y has an effective transition cost of zero because the transition losses for the entire sequence from X through Z are reduced by its presence. In those cases, product Y can be produced in any quantity you wish. The result is another incremental move toward truly lean production.

The third opportunity occurs when you produce D category products on the wheels within wheels. In normal operation, each A, B, or C category product is independently planned and drops out of cycles according to its characteristics. When any member of the category D family of products appears on the reactor, the entire family must be produced in an amount sufficient to remove the entire family from the reactor in subsequent cycles. Often the appropriate production of an entire family of category D products results in production of relatively small quantities for some members of the family. The production of those small quantities is usually quite effective, however, because transitions within the family are typically normal transitions, whereas producing the same product as a transition to or from products outside the family is very expensive.

The transition to or from highly incompatible material makes category D products as a family demonstrate especially high transition costs. By establishing run plans that keep the entire family out of the production cycle, the transitions within the family often have an effective transition cost of zero because the entire cycle of production is better without the entire family of products than it would be if any one of those products appeared more frequently.

Continuous Improvement

As the plastics plant with 200 products and a single big reactor demonstrated, it is often possible to achieve enormous improvements quickly simply by adopting the practice of following a fixed sequence of production. Adding the other elements of a supporting inventory policy and variable volume scheduling significantly enhance those improvements. All of the improvements will occur as a single big event as the result of managerial fiat or professional planning.

That said, there is always a need to produce still more improvement, and FSVV is a great vehicle for enabling continuous improvement. By following a fixed sequence of production, the maximum number of transitions your reactor teams need to manage is greatly reduced. In fact, as category C and D products routinely drop out of the cycle, you will find that the number of the most frequent transitions is reduced still further. As with anything else, this opportunity to focus improvement efforts on a few transitions of great importance significantly improves the ability of your frontline teams to succeed in causing improvement.

If you look carefully at the interrelationships that exist among the elements of FSVV, it becomes apparent that this opportunity for continuous improvement of transition losses immediately propagates throughout the entire system. Each improved transition results in less transition loss for the products involved as well as for the entire sequence of production. As transition losses are reduced, the production of material into inventory to accommodate those losses is also reduced. Better transitions and less accommodating production speed the production cycle. The inventory policy keyed to the cycle length then reduces the amount of material produced into inventory to accommodate the cycle duration, which speeds the cycle yet again. As the cycle becomes increasingly fast, the safety stock that protects against variation in demand within a cycle becomes smaller because less variation can occur during shorter cycles.

Key idea: The relatively large equipment and the continuous operation of many process plants result in chemical inflexibility that is largely unavoidable. Although FSVV will not eliminate those losses inherent to the business, it will promptly give you real improvement and it will enable ongoing continuous improvement. You will have a real opportunity to practice process operations with speed and flexibility that truly begin to approach the lean ideal of producing products in direct relation to current market demand in a way that most process operators could not previously envision.

SUSTAINABILITY OVER TIME

Improvements in performance that result from encouraging people to make extraordinary efforts are rarely sustainable without close supervision and constant management attention. Those effort-derived improvements are also self-limiting because individuals can produce only a certain amount of special effort and only for a limited time.

Alternatively, improvements in performance that are derived from adopting new and better practices are quite often self-sustaining if the new practices are properly adopted. When you deliver good new capabilities to your teams, spontaneous individual efforts usually achieve even further improvement. The pace of change that is possible using this method is rarely limited. In this way, operations that I have managed have sustained a pace of improvement exceeding 16% per year for more than 25 consecutive years. Special effort is not required—only the practice of an improved method at a sustainable level of effort.

For this reason, I normally take great pains to describe the proper way to implement and practice improved methods but say little about sustaining those efforts once they are established. Sustaining FSVV is different. You do need to pay attention to your FSVV practice until it matures.

The "Exception" Problem

Earlier, I observed that the existence of the subway to manage routine personal transportation enabled the effective use of private cars when special transportation was needed. Similarly, the existence of FSVV to manage the routine production schedules enables a plant to manage exceptions effectively when required. There is one major difference: In New York City, each time an individual decides that an exception exists and special transportation is required, that same individual experiences the added cost of using the exceptional mode of transport. In manufacturing plants, the additional cost of providing exceptional service usually is not assigned to the product or the customer that causes it. By enabling plants to manage exceptions to routine production effectively, FSVV often encourages more exceptions.

Fred Smith founded Federal Express in 1973, principally as a service to deliver documents overnight. When the fax machine was developed about 8 years later, some believed that it would replace FedEx for routine, immediate delivery of business documents. Mr. Smith disagreed; he believed that every time people experienced the benefit of speed, they would want even more speed. It turns out that he was right.

Key idea: By enabling exceptions to the schedule, FSVV has often encouraged even more exceptions and therefore has not proven to be as self-sustaining as I would like it to be. In each case where I have implemented FSVV, it has been necessary to revisit the practice routinely to ensure that management is rigorously defending its system and that the frontline team is effectively devoting its efforts toward improvement rather than toward managing exceptions.

In the polypropylene plant where we introduced FSVV in 1987, it was not until 1995 that I was truly confident that the system would continue as a routine part of the business without periodic management assessment of the practice. For this reason, I encourage you to adopt a formal practice of regularly revisiting FSVV—even after you believe that it has been successfully deployed—in order to ensure that it is sustained as you want it to be.

This single concern aside, FSVV is a powerful tool of lean practice that is unique to our industries. It can promptly deliver a large immediate benefit through the big event actions of managers and engineers as well as an ongoing benefit from the autonomous efforts of your frontline teams.

6

Assessment and Improvement of Other Accumulations

INTRODUCTION

The way to be successful in adapting lean practices from one industry or culture to another is to learn the theory of lean enterprise thinking and use it to understand thoroughly how lean is practiced in the original applications. With a theoretical understanding of the examples that exist in other places, you can use the theory to create your own examples and a form of lean practice that will be especially appropriate to your business.

Although the process of directly translating existing lean experiences from mechanical manufacturing to process industries is a valuable use of lean theory, another important element of enterprise thinking is to assess lean practices to determine if some lean tools used in mechanical manufacturing are inappropriate for use in liquid manufacturing. For example, we examined the operating differences between mechanical and liquid manufacturing. We determined that the lean tool *andon,* or "line stop," which is immensely valuable in mechanical manufacturing, is inappropriate for use in process manufacturing because our processes do not stop and start in the same way as those in mechanical manufacturing do. Similarly, many liquid manufacturers, such as petroleum refineries, do not transition their operations among products, so the tools for improving that practice are not applicable. Formal recognition that these popular lean tools are fundamentally inappropriate for our use in a way that is fully consistent with lean theory is a valuable part of the conversation as your team engages with lean practices.

An equivalent opportunity exists to determine if new uses of lean theory might be found to create a special form of lean practice that is uniquely appropriate to your business. Through innovative use of the lean theories, we can go beyond translating what others have done and create new capabilities of our own. Fixed sequence scheduling (discussed in the previous chapter) is an example of a tool that helps us solve the reactive chemistry problems unique to our industry.

In this chapter, we will discuss an extension of the lean theory that inventory exists primarily to accommodate operating problems. The extension that we will adopt is that problems can also be accommodated by a wide variety of other resources, including capital, technology, and technical support. This extension of the basic theory is uniquely appropriate to the technology-intensive and capital-intensive industrial environment of process manufacturing. As a basis for discussing that opportunity, let us very briefly look again at the structural differences between mechanical manufacturing and process manufacturing.

STRUCTURAL DIFFERENCES BETWEEN PROCESS INDUSTRIES AND MECHANICAL MANUFACTURING

It is generally but not always true that process manufacturing is relatively capital intensive and mechanical manufacturing is relatively labor intensive. Using round numbers, in mechanical manufacturing, labor is 70% of the factory cost; material and capital form the other 30%. In process industries, the ratios are often reversed. Labor represents approximately 30% of the cost of manufacturing; capital and materials represent the other 70%. For example, Exxon, a $450 billion process enterprise, employs 80,000 people; Textron, a $13 billion mechanical enterprise, employs 46,000 people. That structural difference creates a real opportunity to explore a potential difference in the focus of improvement activities.

Further, the manufacturing process in mechanical industries often has many small and separate steps with discrete units of labor-intensive production that move between those steps. As a result, mechanical manufacturers have many opportunities for small accumulations of inventory, transport, or other forms of waste between process steps. These accumulations alert operators or managers to the presence of small problems internal to their operation, and frontline teams have many opportunities to recognize small problems in real time and create small event improvements at the frontline of the business. Enabling many people to address many small problems is a powerful force multiplier for any manufacturer and an essential attribute of lean manufacturing.

Because I believe strongly in the enormous benefit of people throughout the business participating in improvement activities, I searched for an equivalent method and benefit that we could enjoy in process industries, which do not employ a lot of labor in small operations and rarely have discontinuities that allow for small local accumulations of material between process steps. Once I understood the implications that arise from the structural differences between mechanical and process industries, I also understood that resources *other than inventory* might naturally accumulate in process manufacturing to accommodate problems that arise that are unique to our industry.

Manufacturing that occurs in small discrete steps will accommodate many small problems by accumulating the resources that are abundant in mechanical operations: work in progress inventory, transport, material handling, and direct labor. In the process industries, we tend to accumulate the resources that are abundant in our industry: technology or engineers, redundant capital equipment, and operational support such as maintenance or laboratory technicians. Searching for work in progress, material handling, or direct labor internal to our operations is an unlikely source for identifying small improvement opportunities in the process industries. Searching for other small accumulations or wastes that are abundant in our industry has real value.

Small Accumulations in Process Industries

Most of these small problems and small accumulations are amenable to resolution by autonomous action at the frontline in the same way that small accumulations of work in progress or direct labor enable frontline improvement in the mechanical industries. This represents a new source of improvement for process manufacturers, which can be added to more traditional improvement methods. Mechanical manufacturers may find that they also have their own form of these alternate accumulations in the same way that we have found our version of their problems. In any event, the expanded use of lean enterprise thinking that we will use for the purposes of this discussion is a simple generalization of the original lean theory that inventory hides problems.

Key idea: The resources that accumulate to accommodate problems in manufacturing are those that are particular to the operation. Inventory can be one of these accommodating resources but it is not the only one. Other resources can accumulate for the same purpose with the same effect. In the process industries, small opportunities for improvement internal to the operation are more likely to manifest in one of these alternate ways than as inventory.

In the process industry, we can use this general theory to engage people throughout the business. In exactly the same way that mechanical manufacturers engage people to identify and resolve small problems using inventory as the indicator of a problem, we can achieve the same outcome by using other accumulations as the indicator. We are discussing a straightforward extension of the original theory of lean enterprises; the enabling technologies for resolving the underlying problems are described elsewhere or are unique to the situation. Therefore, this chapter will consist of four examples to illustrate this concept and give you some ideas of where to start looking for similar opportunities.

Case Study: Testing for a Problem That No Longer Exists (Excess Technical Support)

When we began to deploy the technologies of lean in Baytown, we simultaneously began a serious effort to engage all employees to help us practice improvement. It helped to have the benefit of prior lean experiences, including experience in liquid manufacturing (see Chapter 1). That experience immediately allowed us to give many engaged people throughout the business the general version of lean theory and invite them to look within their areas of responsibility for any form of accumulated resources that were "unnatural."

Note: "Unnatural accumulation" means any buildup of any type of resource for which there is no obvious operating need.

The likely accumulations that are uniquely appropriate to process operations, such as technology, capital, or manufacturing support, were offered as thought starters for that effort. The expectation that we shared with the staff was that an unnatural accumulation of resources was likely to be an indication of an unresolved problem and that the unresolved problem likely represented an opportunity for improvement. Very quickly, one of our young supervisors found exactly the sort of thing for which we were looking when he questioned a resource that had been in place for so long that most people with more experience did not even notice it. Certainly, the more experienced people did not regard it as unnatural.

Isopropyl alcohol (IPA) is a well-established and fundamentally simple product that we had been producing for decades. We never had a problem producing it and our customers never had a quality problem with it. We surrounded the production of IPA with extremely robust and reliable quality systems that consistently worked extremely well. Further, we used dedicated equipment to load the product into dedicated rail cars. Those rail cars then moved directly to the customers, who used dedicated lines to empty the cars, which were returned directly to us, sealed and empty. There was no realistic possibility that the product could become contaminated or otherwise degraded in the shipping process.

Why then, this shipping team leader wanted to know, did we employ a laboratory technician all day and all night, every day and every night, to test for contamination after the good product had been loaded into a clean and dedicated tank car? Why did we hold tank cars at the loading rack until the analysis returned from the lab before releasing the product for delivery to the customer?

Once he raised the question, the answer was very interesting. First, he determined that the analytical procedure was indeed an accumulation of technical resources that had real cost. An Exxon laboratory technician with salary, benefits, overtime, uniforms, and all the other costs of employment has a cash cost of more than $100,000 each year. Having a technician continuously available all day every day required that we have a minimum of five individual technicians. We found that we were paying at least $500,000 each year to test for contamination problems that had not occurred even once within organizational memory or within any records that were available to us. In fact, we believed that such contamination could not occur. Further, as with other resources that accumulate, there are always other costs. For example, lab techs need laboratories, lab supplies, and lab supervisors. The $500,000 estimated cost of this accumulation was certain to be low.

Upon investigation, it was discovered that the test for contamination in the rail cars had been instituted at the very beginning of IPA production, more than 50 years earlier, when we shipped IPA using a fleet of shared rail cars that routinely carried many different products. With shared rail cars, the possibility of contaminating the IPA product as it was loaded for shipment was very real. Exactly in line with lean theory, there was an operational problem and we added the resource of a laboratory technician to accommodate the problem by ensuring that if contamination did occur, it would be detected and contained within our plant and would never reach our customers.

By the early 1990s, when this investigation occurred, we had been using dedicated rail cars and dedicated equipment to move IPA to our customers for more than 10 years. However, no one had thought to stop testing for the contamination. We had accommodated the potential problem so thoroughly that no one remembered why we were doing the testing and no one noticed that, due to other changes, the problem for which we were testing no longer existed. Needless to say, after we came to this understanding, we promptly discussed the situation with our customers, who agreed that we could stop testing.

This is a perfect example of the value of teaching frontline people about lean enterprise thinking and the general theory of the accumulation of resources. Exxon Baytown is an enormous plant situated on over 3,500 acres and employing thousands of people. This particular "accumulation of resources" had been in place for longer than most of the people working at the plant. The appearance of the laboratory technician at the loading rack each day was uneventful. A single person arrived a few times during each shift to take a sample out of rail cars, return it to the lab, and later release the cars for shipment. This same thing also happened routinely for many other products.

If a manager or engineer noticed this activity at all, he or she was simply pleased that the lab results continued to confirm that everything was good. As evidenced by the fact that the problem had existed without resolution for more than 10 years, it is unlikely that plant management would ever have noticed this as something that needed to be studied and eliminated. Identifying and resolving this problem required someone at the frontline who knew what to look for, recognized what he was seeing, and had the capability to initiate action to get it resolved. Once the problem was recognized, resolving it required careful management of customer interactions, but really did not require any special capabilities or costs.

Case Study: Maintenance Malfunction (Excess Capital Equipment)

In continuous process chemical operations, it is common to have spare pumps. Although heavy-duty, high-volume chemical pumps are often made of an exotic corrosion-resistant material and are expensive, they are a relatively small part of the assets of a large chemical plant. Owning spare pumps is very inexpensive compared to the cost of allowing a large continuous operation to stop production as the result of a pump failure. As you walk around a large chemical plant, it is common to see pump slabs where it is obvious that there are two pumps for the same service: one in operation and one standing by as a spare.

When a pump that is in service fails or is otherwise taken offline, the spare pump is brought into service and the plant continues to operate normally while the original pump is repaired. This practice is so common that by using remotely operated valves and switches, the change between pumps in service is often achieved from the control house without sending a technician into the field. In some places, the process control computer recognizes that a pump has failed or is in distress and converts to the spare pump without even requiring the intervention of the console operator.

Although it is common to see pump stands with a pump in service and a spare pump available, it is much less common to see a pump stand with three pumps ready for the same service: a pump in service, a spare pump, and a third pump as backup for the spare. For engineers to create such a situation, it must be reasonably likely that the spare pump, when activated, might fail before the original pump can be repaired and returned to service. Because a chemical pump that is well designed for its service normally lasts for months or years, the presence of three pumps was a clear indicator that something abnormal was being accommodated.

After we launched our lean activities, one of our employees questioned the fact that in some places we had three pumps for the same service. One spare seemed to represent a normal contingency against failure, but two spares must represent an unnatural accumulation. What problem was it covering?

We were greatly embarrassed when we investigated this unnatural accumulation of capital resources. First, we found that, indeed, it was not only possible but also quite common for the spare pump to fail before the original pump was returned to service. That was very odd because, once a pump began to operate routinely, it often ran for many months before failing. Yet once the primary pump failed, it was common for the spare pump to fail soon after.

It turned out that although we were doing a lot of preventive maintenance on the spare pumps to ensure that they would be available for service, we were doing it incorrectly. In this particular service, the start-up viscosity to get the material moving was extremely high. Once the material was in motion, it moved well enough, but it was hard to get it moving.

When we designed our preventive maintenance program for these spare pumps, we applied industry standard maintenance practices instead of designing maintenance routines that were particularly suited to the exact operating context for this equipment in this service. Included in those standard maintenance practices was a commonly taken step to prevent idle electric motors from getting a “flat spot” on the rotors. This is a common problem for large electric motors that are not in service for long periods. The preventive maintenance practice to prevent flattening of rotors is to rotate the drive shafts of idle motors periodically. For an installed motor, the technicians had adopted an easy way to achieve this rotation, which was to turn the motor on and let it run for a few minutes each week.

The practice of rotating electric motors that are not in service is a good idea if done correctly. In this situation, however, the motors were not in storage; they were installed and connected to pumps. By starting the motors, we regularly exposed them to the extreme stress of overcoming the high start-up viscosity of the material in the pump. Each time that we started the motors, they experienced the most severe operation that they would ever experience. By frequently starting and stopping the motors, we were doing the one thing that was most likely to cause the motors to fail when we needed them.

These maintenance practices seemed like a good idea, but once we assessed the problem, we stopped doing them in the same way. From that point forward, we continued to rotate the motor shafts as part of our preventive maintenance plan, but we did it by hand rather than by starting the motor. With that modest change, the problem of premature pump failure in this application was eliminated.

Key idea: In 1991, Exxon had not yet developed appropriate maintenance taxonomy and reporting practices and our poor records and poor use of nomenclature hurt us. The equipment failures were regularly reported as "pump failures," although the *pumps* were fine. The pump *motors* were failing. Our reliability engineers did not review the maintenance program for the motors because only the maintenance technicians knew that the motors were the problem. Our reliability engineers also did not know that the technicians were rotating the motor shafts by starting the motors. In a way that is quite common with improvements at the frontline, uncovering and correcting this situation led us to a long series of other improvements.

Because of this small change in maintenance practices and several other changes that were functionally similar, we were able to redeploy many high-cost chemical pumps to new service rather than purchase additional pumps. We were also able to eliminate the cost of rebuilding motors that should never have failed. Of even greater importance, when we had two spare pumps, we applied our improper maintenance practices equally to both spares. Therefore, even with two spare pumps, we periodically experienced an event where a main pump failed and both spares also promptly failed. When this happened, we had to stop production at a very high cost. By recognizing and correcting the relatively small problem identified by examining an unnatural accumulation of spare pumps, we avoided the bigger problem of bringing down a production line.

Note: Maintenance that does harm rather than good is far more common than you might suspect. For example, at Suncor, we routinely experience erosion in our pipes, but that erosion has a predictable pattern of appearing on the larger radius of pipe bends and elbows where the flowing material impinges as it changes direction. One of our erosion monitoring technicians noticed that we were replacing some piping because of erosion on the smaller radius of the bends. There is no operating reason for that to occur. Upon investigation, we learned that our maintenance team was inadvertently causing this damage when they aggressively "pigged" the furnaces to remove accumulations of coke on the insides of the pipes. After that discovery, we managed the size of the pigs and the frequency of pigging much more carefully and eliminated this self-inflicted source of pipe damage. As in the Exxon situation, once we identified this problem in one location, we were able promptly to identify and resolve similar problems in other locations. Frontline teams, who are aware of this potential and watching for odd maintenance results, are a good source of this intelligence and a lot of improvement will flow from it.

Key idea: At both Exxon and Suncor, once the maintenance problem was recognized by a frontline person and the reason for it understood, it could be resolved with high value and little cost. The opportunity to create big improvement with small cost is exactly what we are looking for as we engage people to help us implement lean or any other improvement tool.

Case Study: Solving Problems Before They Occur (Preventing Excess Capital)

In some situations, the unnatural accumulation of resources has not yet occurred. By recognizing that such an accumulation is about to occur, it is possible to jump directly to fixing the problem without experiencing it.

FIGURE 6.1
A bale of synthetic rubber.

In Chapter 4, we described the use of an extruder to "finish" polyethylene and polypropylene. In that case, the newly made polymer leaves the reactor in the form of low-density crystalline granules. For ease of shipment and for improved processing by our customers, the extruder makes the polymer crystals into dense pellets. We do a similar thing in the manufacture of synthetic rubber. As it leaves the reactor, synthetic rubber has the form of a low-density "crumb" that looks surprisingly like popcorn. Before shipping rubber to the customers, we form it into dense rectangular blocks that have an appearance similar to small white hay bales, as shown in Figure 6.1. In fact, those blocks are called bales of rubber. The bales are packaged for shipment by stacking them into large cardboard boxes. One of our rubber plants purchased an automated machine that would erect these boxes from the flat cardboard preforms that we received from our box supplier.

As we gained experience with it, we learned that the box machine tended to jam when the factory air pressure dropped just a little below its normal operating level. Having learned that, the operating crew in the packaging area proposed that we purchase an air compressor dedicated to the box-making machine so that it could operate independently from the factory air supply. This proposal assumed that the box machine was fine and that the operating problem was caused by the natural variation in the factory air supply. The members of the crew did not intend to fix the box machine so that it would work with the existing factory air and they did not intend to remove the variation from the factory air. They simply intended to add a new source of compressed air.

As with most resources added to accommodate unresolved problems, the compressor would have introduced new costs for spare parts, operation, and maintenance. The team might have gotten its compressor nevertheless, but made the classic mistake of asking for belts and suspenders. The team not only wanted its own compressor, but also a spare compressor in case problems occurred. That request had the unmistakable appearance of accumulating capital resources to cover an unresolved problem.

When we sent that team back to review the situation, it learned still more about how its box machine operated. According to the design specifications of the air-actuated valves that were failing to operate, they should have operated normally at pressures well below the lowest pressure that the machine was experiencing. Although it was clear that when the box machine jammed, the events correlated perfectly with episodes of low factory air pressure, the team concluded that because the valves should not have failed at that pressure, some other, more fundamental cause must be responsible for the failures.

Upon still further investigation, it became obvious that the machine was simply plumbed badly. In most cases, the air moved inside the machine through quarter-inch tubing. As a result, the airflow through the machine was so inefficient that the valves farthest removed from the air inlet were in jeopardy of failing even at normal factory air pressure. When the factory air pressure dropped just a little, the valves failed. The manufacturer of this box machine normally produced machines that erected small boxes such as those used for groceries or cola bottles. Our machine was the largest machine that this manufacturer had ever produced and as it designed our large machine, it did not properly scale up the plumbing designs.

At the end of the day, the team operating the box machine did nothing more than have the machine replumbed using half-inch pipe rather than quarter-inch tube. It cost some maintenance money to do that, but the cost was nothing in comparison to buying and operating new air compressors.

In this case, although some management intervention took place, it should be clear that the frontline team recognized the problem initially and ultimately found the low-cost resolution. Management participation was limited to recognizing that because of an unresolved operating problem, there was *about to be* an unnatural accumulation of resources and then sending the team back for further investigation. As with the other examples, by looking for the problem that was hiding under the accumulation of resources—in this case a *potential* accumulation of resources—we found a problem that was relatively easy to resolve.

There is nothing at all improper about management intervening to assist teams at the frontline as they learn about the proper practice of small event improvement. A normal function for management is to teach people the new system and help them as they learn to use it well. In this situation, the lesson is basic: Small event improvement is always best when it is conducted within the resources naturally available to the teams. Modest maintenance expenditures fit that model well. Capital projects do not.

Case Study: Leading the Improvement Process (Defining Excess Maintenance)

As a result of disciplined practice of manufacturing improvement at Exxon Baytown, we routinely experienced equipment performance that far exceeded industry standards. Over time, we began to wonder why we needed maintenance technicians on nights and weekends. Why could we not do all or nearly all of our maintenance during the normal work-week? At the same time that we asked that question, we recognized that the need to operate the plant was far greater than the need to reduce maintenance costs. We never forgot that rule number one of lean practice teaches us that we want to remove resources in a way that lets us identify problems or sustain an improvement, but not in a way that causes us to experience problems.

With that in mind, management arbitrarily defined evening and weekend maintenance as an unnecessary accumulation. We fully accepted that we would continue to do maintenance on nights and weekends as needed, but that maintenance during those periods would be an exception rather than the norm. In that manner, we greatly reduced the maintenance staff available on nights and weekends by assigning them to work during the normal weekdays.

Only a small maintenance staff was routinely present at night. Normally, its members performed a limited amount of planned work, but they were available as a contingency in the event that we needed them. It became our policy, however, that unplanned maintenance performed on nights and weekends would occur only if it were needed to react to an unforeseeable operating problem. Unplanned maintenance at night and on the weekend was permitted only if the operating superintendent declared the situation to be a reportable incident that would be reviewed by senior management the following day.

In that environment of arbitrarily limited maintenance resources, operating supervisors began to take a close and discriminating look at events that previously would routinely have resulted in off-shift maintenance. As expected, we found a wide variety of modest but recurrent problems that had been accommodated by abundant maintenance resources instead of solved.

In a very orderly way, we scheduled those problems for planned resolution. Some were harder to resolve than others, but all were well within our normal maintenance capabilities once we had identified them and created a focused effort to resolve them. It is important to note that these problems did not exist only on nights and weekends. By declaring night and weekend maintenance to be unnatural, we identified and resolved a bundle of small problems that had previously cost us money and production all day, every day.

As described in Chapter 1, during the 6-year period from year-end 1991 through year-end 1997, Baytown improved productivity by 16% each year, resulting in a plant with double the productivity and double the profitability that it had previously enjoyed. Over that full period, more than half of that increase in productivity resulted from increased capacity obtained by improved operation of existing assets. In 1995, our fourth year of improving capacity and productivity at this pace, we needed a new source of opportunities for improvement in order to maintain our rate of progress.

This practice of arbitrarily defining off-shift maintenance to be an unnatural accumulation of resources served us well as a way to identify new opportunities for improvement. This practice was precisely in conformance with lean enterprise thinking. By withdrawing the resources that previously accommodated many small operating problems, we required that the problems receive the attention required for resolution.

Asset performance continued to improve through the autonomous actions of many people who had a new focus on paying attention to the details. Although we did withdraw resources in order to identify unrecognized maintenance problems, we retained the capability to respond promptly to those problems as they occurred until we permanently eliminated them. *We never withdrew resources in a way that forced us to experience problems that we had not yet solved.*

Case Study: Defining a Scarce Resource

We had a similar experience at Suncor. As our capacity to transform bituminous sand into synthetic crude oil grew, our fleet of mining equipment to recover the oil sand from the ground also grew. Oil sand mining equipment is among the largest land-mobile equipment in the world, with shovels that extract 100 tons of ore at a time and trucks that transport 400 tons per load. As our fleet grew, our need for heavy equipment technicians (HETs) also grew and HETs actually became such a scarce resource that maintenance of mining equipment became a rate-limiting step in our growth. This was true not only for us but also for the entire industry. As a result, we arbitrarily defined HETs as a scarce resource and we began systematically to assess every task that required an HET to take any action that was not a normal part of maintaining our fleet.

In this way, we found just what the lean theory predicted we would find; an unusual amount of our scarce resource was consumed by a number of small problems. Many of these problems were minor defects with the equipment the HETs used to service the trucks and shovels, the garage facility where they worked, or the vehicles they used to access equipment that failed in the field. We added a crew of millwrights and electricians in the HET shop to keep the equipment and facility in repair and we added an auto mechanic to keep their vehicles in repair. With this reduction in distractions, the HETs became much more productive at keeping our mining equipment in repair.

Note: Use of the SMED and other lean tools to improve the execution of routine HET mechanical tasks, such as accelerating the change-out of truck engines described in Chapter 4, also helped.

Key idea: Adding millwrights, electricians, and auto mechanics does sound like it represents an accumulation that itself covers a problem. An alternative view is that by deploying a standard maintenance crew to provide normal support for the facility, we were able to redeploy a scarce and more valuable resource that had previously accommodated unresolved maintenance problems. As you deploy lean theory and practices in your plant, it is necessary to understand what problem you are trying to solve.

There is a lot of value, especially in process plants, in looking for accumulations of resources other than inventory that are likely hiding problems unique to each individual operation. This is one of our best paths toward enabling people at the frontline in our industry to practice the continuous small event improvements that mechanical manufacturers enjoy from more traditional lean practice.

7

Statistical Quality Improvement

INTRODUCTION

The initial development of the Toyota production system and the related technologies of lean manufacturing in Japan followed the introduction of statistical process control (SPC) to Japanese industry so closely that they are often considered contemporaneous. Certainly, they are completely aligned and the differences between the two approaches are more complementary than conflicting. This is one reason for the recent growth in Western industry of combined programs, sometimes called "lean six sigma." This is not a new concept. It is more a reunification of historic companions. Lean manufacturing and statistical process improvement are made for one another.

THE POWER OF STATISTICAL QUALITY COMBINED WITH LEAN MANUFACTURING

As everyone who participated in Western manufacturing during the mid-1970s knows, the early adoption of these two practices in many Japanese industries resulted in improved product quality at the same time that manufacturing costs went down. The devastating competitive effect was that General Motors and others found themselves competing with companies such as Toyota, who became quickly able to deliver better products at lower prices. That is essentially the same thing competitors of Gilbarco and Exxon Butyl Polymers experienced (see Chapter 1). General Motors knew what Toyota was doing, but could not understand how to match that performance. As we adopt lean in the process industries, we will also want to have the natural companion capabilities that come from statistical process control.

Just as lean principles have changed the paradigm of traditional manufacturing practices, SPC has changed the conceptual model of product quality and process consistency. Statistical process control shifts the emphasis away from the lagging indicator of postproduction inspection toward the leading indicator of assessing and controlling the inherent variation of manufacturing processes. The new theoretical construct is that, when a manufacturing process is inherently capable of successfully producing a product and that process is controlled to operate without variation from that capability, the quality of the product will assuredly be good. In this way, product quality can be controlled by proactive measures to improve the process and control variation rather than through reactive responses to accommodate poor results.

Thus, SPC has two essential components. The first is assessing and improving the inherent capability of the process and the second is operating the process without variation from that capability. Statistical process control provides engineers a proactive opportunity to improve the process and it provides teams at the frontline a proactive opportunity to operate the process without variation. Like lean, SPC enables everyone to make a meaningful contribution toward achieving the lean ideal of perfect production.

Statistical Methods in the Process Industries

Adaptation of statistical methods from mechanical manufacturing for use in the process industries is straightforward. With the exception of some modest computational differences to account for liquid effects common in process manufacturing and virtually unknown in discrete production, such as autocorrelation of material commingled within a single reactor, statistical methods are applied in very similar ways in both industries. For this reason, Exxon Mobil and other leading process manufacturers have been widely practicing SPC in liquid operations for more than 30 years. Indeed, because liquid manufacturers have special problems with process stability and consistency, it is clear that, once again, we have potentially more to gain from these methods than mechanical manufacturers do.

Case Study: Managing Variation

The fundamental concept and benefit of assessing and controlling "variation" as a means of managing the performance of manufacturing processes are not always self-evident. Therefore, an example is often useful to introduce this topic. One of the best examples of controlling variation came early in my career while I was leading a team that made instrument panel safety pads for General Motors. For reasons of styling and appearance, the instrument panel pads were covered using vacuum-formed vinyl sheets. In the original design, the flat sheet of vinyl prior to vacuum forming was specified as 0.040 inches thick with a tolerance of plus or minus 0.004 inches.

The engineering purpose for this specification was to accommodate the stretching and thinning that occurred as the flat sheets were vacuum formed to the shape of the instrument panel. By starting with a target thickness of 0.040 inches, we believed we could ensure that the final thickness in all locations on the instrument panel would be greater than 0.020 inches. However, variation in the thickness of the starting sheet had a great impact on the production results.

The relatively thicker portions of the sheet heated more slowly and therefore exited from the heating oven at a lower temperature than the thinner portions. During the forming operation, the portions of the sheet that were both thicker and cooler were naturally less elastic than the thinner and hotter portions. Therefore, the portions of the sheet that were originally thicker stretched far less than the portions that were originally thinner. In fact, the existence of relatively cool and relatively inelastic places on the sheet caused the thinner and hotter places to stretch even more than they otherwise would have.

We went through several iterations of reducing the thickness variation in the flat vinyl sheet. Each time that we improved the variation in the sheet, the performance difference in heating, stretching, and forming also improved. By the time we were able to produce vinyl sheets with a thickness variation of less than 0.001 inch, the sheets formed into a finished product with a uniform thickness. This new consistency in our process allowed us to reduce the target starting thickness for the flat sheet before forming from 0.040 to 0.030 inches and still meet the target for producing formed instrument panel covers with a minimum thickness of 0.020 inches.

In fact, we met the minimum thickness target for the finished product more frequently with a thinner and more consistent starting material than we did previously with a thicker material. We did not change any characteristic of the material or of the product; however, by improving the consistency of the process, we were able to reduce total material consumption by 25%. Further, the now more consistent material also performed better in the field. The instrument panel pad in its final form continued to benefit from the lack of variation in thickness as it responded to the environmental heating and cooling experienced by the cars during their lifetimes.

Key idea: Improved consistency, or reduced variation, and nothing more allowed a significant reduction in cost accompanied by a significant improvement in product quality.

The existence of natural process variation has been long recognized in industrial quality practices. It is apparent in product specifications that provide for a tolerance around the optimum measurement. For example, the vinyl sheet was specified to have a thickness of 0.040 inches plus or minus 0.004 inches. This tolerance implied that the product designer anticipated a manufacturing process that would produce up to 10% natural variation on either side of the optimum result.

Unfortunately, prior to SPC there was no practical method to assess and manage variation as an attribute of individual production systems. The designer of instrument panel pads had no specific knowledge of the natural variation of our sheet extruders. Ten percent variation on either side of the optimum seemed like a reasonable target. Still more unfortunately, when natural variation was approached or considered to be an unmanaged and perhaps unmanageable attribute of production, it did not invite the focused improvement that is possible today.

Basic Statistical Concepts

The most basic element in the industrial practice of statistical methods is process capability. The origin of this concept is generally credited to W. Edwards Deming. Deming's essential idea is that manufacturing processes are inherently variable, but variation in the process is both *measurable* and *predictable;* more importantly, once known, natural variation is *manageable.* In this way, SPC gives manufacturing leaders the ability to control an attribute of our business that we previously only experienced.

Practically speaking, in manufacturing, the importance of natural process variation becomes evident when the statistically predicted output of a production process is compared to the specification of the products to be produced by that process. For example, variation in thickness measured in thousandths of an inch, which was very important in the production of vinyl sheets, probably has no importance at all in the production of marine anchor chains. Considered in this way, it becomes clear that the control of process variation that is appropriate to one product is likely to be inappropriate to the other. *The value of managing variation in the production process is product specific.*

Key idea: Deming taught us that the inherent consistency of industrial processes could be analyzed statistically to create a mathematical model that will accurately predict the natural performance of each process. By comparing the natural performance of a process with the requirements of the products produced by that process, *it is possible to describe precisely the capability of the process to produce the product successfully.* Once the inherent capability of a process is known in detail, *it is possible to improve the operating factors that have an impact on that capability and it is possible to manage the operation of the process effectively to sustain that capability.* That is the essence of statistical process control.

Six Sigma

Because this practice depends upon statistical analysis, mathematical notation is used to compare process variation with product requirements. Statistical practice reports variation around a mean value in terms of standard deviations, designated with the Greek letter sigma (σ). Statistically, nearly all of the natural variation of a process (99.9997%) is contained within three standard deviations on both sides of the mean value. Thus, nearly all of the natural variation of a process will be contained within a total range of six standard deviations or within a total range of six sigma.

A process that is adjusted so that the central or mean value of the output is aligned with the optimum value of the product specification is described as a "centered process." When a centered process also has a range of statistically predicted natural variation that is the same as the tolerance range of the product specification, all of the natural output of that process is contained within the product specification. The process is naturally and routinely capable of satisfactorily producing that product. According to recent industry practice, this process is a "capable" process and the situation is described as "six-sigma quality."

In that situation, the primary source of defective products will come from conditions or events that are not normal to the process. These abnormal events are described as "special cause" defects. The task of improving *the fundamental process* is to create a process that is naturally capable of producing specific products. The task of improving *process operations* is to protect the process from special causes that disrupt the natural capability of the process.

Using Deming's methods, it is possible to create a mathematical model of process operations that predicts with great precision the capability of each manufacturing process to produce each product. Further, once the statistical model of process variation exists, it is possible by experiment or experience to identify the sources within each process that produce the variation. By proactively identifying and removing or improving the sources of natural variation, it is possible to make the process inherently more capable.

Furthermore, once the statistical model of the process exists, it is possible to identify periods when the process is not performing according to its natural capability. Identification of such periods makes it possible to know with precision when nonsystemic or special cause variation is influencing the performance. Thus, special cause variation can also be proactively identified and removed as it occurs. As special causes of variation are removed, the *natural capability* of the process *is not* improved, but the *immediate performance* of the process *can be* improved.

The statistical model of the process created by SPC practice thus enables two powerful new approaches to process and performance improvement that could not otherwise exist. Through targeted experimentation supported by statistical analysis, it is possible to identify the sources and impact of natural variation within the process. By removing natural variation, it becomes possible to improve the inherent capability and the *long-term performance* of the process. Similarly, by continuous comparison of actual performance to the statistically predicted performance, it becomes possible for operators to identify in real time when special causes are affecting the production. With nearly perfect correlation between cause and effect, it is possible to identify and eliminate special causes of variation in a way that improves *immediate performance.* Permanent removal of recurrent special causes can also improve the effective natural capability of the process.

This understanding and action make proactive quality improvement possible. Previously, we inspected the finished product and reactively adjusted the process. In statistical practice, we can proactively analyze and improve the process and we can continuously monitor process performance to protect it from abnormal events.

Process Improvement before Statistical Analysis

Very few manufacturing processes are completely incapable of producing good products. Most industrial processes produce quite a few good products. In the nature of random variation, those "pretty good" processes have periods of great performance and periods of poor performance—all within their natural capabilities. As a result, prior to SPC, when we operated a process that generally produced 98% good products, we were happy when the success rate was periodically 100% and sad when it periodically dropped to 96%. We did not recognize that both the good and bad results coexisted within the natural capabilities of the process.

For this reason, we did not have a well-structured approach to process improvement. Engineers and managers generally assumed that the existence of good periods demonstrated that the process we had created was a good process, and therefore the bad periods must be the result of something other than our good process. As a result, for processes that periodically but not always produced great performance or, as we would now describe them, processes that were nearly capable, most improvement efforts prior to Deming focused on searching for and removing special cause problems.

Much of the search to identify special causes focused on operator performance, which often had nothing at all to do with the natural variation of the process. Second only to operator performance, improvement efforts depended largely on engineers or operators adjusting the process to recenter the mean value of the output. Without understanding that the variation we were seeking to affect was inherent to the process, seeking improved operator performance often had no impact and recentering the range often made the situation worse. We periodically found and resolved a true special cause or a source of natural variation, but these activities were unstructured and infrequent.

As a result, improvement prior to SPC was much slower and much less successful than is now acceptable. Improvement efforts that reactively focus only on operator performance, process adjustment, and special cause elimination are unlikely to enable manufacturing to approach the state of perfect (or defect-free) operations because they ineffectively address only one of two possible sources of defective products: special causes. Improvement of the inherent capability of the process is rare in that situation.

Process Improvement Using Statistical Analysis

Because each process has an inherent capability that can be precisely measured and meaningfully related to the specifications of the product to be produced, SPC enables us to create a useful analytical tool for structured and widespread process improvement. In simple terms, this tool is a statistical model that describes the expected natural performance of the process. By comparing the statistically predicted performance of a process to the specification limits for the products to be produced, it is possible to predict precisely the extent of success that each process will have in routinely manufacturing each product. To the extent that the process capability demonstrated by this analysis is not sufficient to meet the needs of the product, more detailed experimentation and analysis can be conducted to identify and remove the sources of variation inherent in the process so that the inherent capability increases.

Case Study: Using a Predictive Model

In Suncor's extraction area, we routinely experienced periods when the process did not satisfactorily separate the sand from the bitumen. As a result, excessive sand was delivered to the upgrader along with the bitumen. This caused erosion in our pipes and furnaces. For a long time, we lacked a predictive model of the extraction process. Bitumen quality was determined by an inspection conducted after the bitumen was produced. In a classical approach to product quality, when we found that the bitumen contained too much sand, we adjusted the operation of the extraction unit, typically by running it more slowly. Operating slowly improved product quality, but it constrained capacity.

After we adopted statistical methods, we found that the extraction unit was inherently incapable of routinely producing bitumen without sand at full production rates. The natural variation of our process periodically exceeded the range of the product specification. Knowing this, we conducted a more detailed analysis and determined that the source of this variation in product quality was determined largely by the consistency with which we introduced feed into the separation cells. When the rate of feed into the cells surged up, the amount of sand in the bitumen surged up as well.

We had thought the separation cells to be the site where the discrete operations of our mine transformed into the continuous processing of synthetic crude oil production. However, this analysis taught us that the change to continuous flow had to occur as material entered the separation process, rather than within that process. With that knowledge, we were able to improve the inherent capability of the process to operate at full rates while producing satisfactory product.

Operational Improvement with Statistical Analysis

In addition to providing a tool for improving the process, the statistical model of process capability can be utilized as a tool for ongoing assessment and improvement of real-time performance. By comparing current performance to the predicted normal performance, it is possible for frontline teams to unambiguously identify when the process is not performing as expected. At that point, the teams also know with some certainty that the process is subject to a nonsystemic or special cause of variation. Because this comparison and assessment are happening in real time, the frontline teams also know with some certainty that the special cause they are seeking is something that has *just* changed or that *is happening* at that moment.

With the ability to identify closely when nonsystemic variation is occurring, the search for the source of that variation is much more successful. In fact, recognizing that SPC identifies when the process is not performing as expected, *independently of the inherent capability of the process,* it is possible for operators to separate special causes of variation precisely from natural variation in the process. This powerful analytical ability is available for processes that are inherently capable as well as those that are inherently incapable. Incapable processes will naturally produce some quantity of defective products, but so long as it is clear that those defects are the result of a process that is operating as expected, the operators do not need to adjust the process or search for a special cause that likely does not exist. In the search for opportunities to drive improvement, there is amazingly powerful value in knowing with certainty whether the process is or is not performing as expected. That knowledge is not possible without SPC.

Key idea: Statistical analysis enables several different improvement paths to operate simultaneously. Once we have a statistical model that accurately predicts the natural performance of the process, we can work to analyze and improve the sources of variation inherent to the process further. We can also begin to identify independently when any process is not behaving as it should and either adjust it or search for the nonsystemic cause of that unnatural variation.

STATISTICAL MODELS OF PROCESS PERFORMANCE

What exactly is a statistical model of predicted process performance? *The natural performance of a manufacturing process is the performance achieved when the process operates without adjustment.* To assess this performance, the process is set up normally, stabilized, and then allowed to operate while the results are sampled periodically. The sampling during this period measures the inherent performance of the process, rather than the skill of the operator in adjusting the process during operation. Using statistical methods ensures that the sample is meaningful, but limits the amount of time the process must run without adjustment in order to gather adequate data.

Key idea: Measuring the unadjusted performance of chemical process equipment is a severe standard. It is often a real wake-up call for process engineers who never before recognized the extent to which the processes they designed were dependent upon constant attention and adjustment from the frontline teams.

In all processes, data gathered from sampling the unadjusted output enable two important statistical calculations. First, the analysis will compute the mean, average, or central value of the result that the process naturally produces. This is a measure of the ability to stabilize the process on a selected target value during the initial setup. Next, the analysis will compute the range of output values around the mean value and the expected distribution of values within that range. This is a measure of the natural variation inherent to the process. Taken together, these two results allow you to create a mathematical model that accurately predicts the natural performance of the process.

It is equally possible to gather the data required to create a model of your process by using the brute force and awkwardness approach of taking a great many measurements over an extended period. If you gather enough data for a sufficient period, you will produce from long experience nearly the same model of expected process behavior without statistical analysis. That experiential model would allow you to understand process capability and react to it in the same manner as a statistical model. Without knowing that we were doing this, many of us had gathered large quantities of data prior to having the SPC tool and used that data to produce good process improvement. That is essentially what we did to solve the vinyl sheet problem described earlier.

There are three problems with the brute force method. First, gathering massive amounts of data is painfully difficult. Second, most operators will not leave the process without adjustment for sufficiently long periods to enable a long-term study of performance. When you gather data from an adjusted process, the data reflect the natural variation of the process as well as the special variation caused by operator adjustment. Third, when experiments are conducted or changes are made to improve the process, you need to gather the data painstakingly once again in order to assess the results and sustain the benefit.

I mention this nonmathematical alternative approach to assure you that statistical analysis does not produce an inexplicable outcome. Statistical analysis simply makes it easy to do something that you could also do in another, more difficult way. The power of SPC is similar to other attributes of lean that we have been discussing in that it enables many people to do routinely and easily what previously was limited to a few specialists or perhaps was not done at all due to the difficulty of execution.

Often, using a statistical sampling approach, a workable process model can be created in a relatively short time. Routine maintenance of the model, once it exists, to reflect process improvements or other changes is also quite easy using statistical methods. The statistical methods for gathering the data and performing the analysis are thoroughly documented; therefore, I will not reproduce that discussion here.

In most plants, the statistical models are produced by professional statisticians or by specially trained people such as six-sigma black belts or green belts. For most people, the much more interesting part of SPC is what you do once you have a mathematical description of the capabilities of your process. Most engineers and frontline teams can make productive use of these models.

Using Statistical Analysis: The Process Capability Index

By comparing the natural performance of the process, as described by the statistical model, to the specification of the products to be produced, it is possible to determine with good precision the capability of each process to produce each product. The result of that comparison is called the "process capability index," which is always a very specific relative value comparing the natural output of one process to the specification of one product.

A capability index of 1.0 indicates a process with six-sigma quality—that is, a process where the natural variation during centered operation precisely matches the specification range of the product. Processes with a capability index of 1.0 will produce defective product only if they become uncentered or if a special cause disrupts the natural capability of the process. Higher values for the process capability index indicate processes that are more robust because they have excess capability. Processes with excess capability may be able to tolerate some special cause variation and still produce good products because even the excessive variation arising from the special cause may still be contained within the product specification. Capability indexes less than 1.0 indicate a process that is naturally incapable and will always naturally produce some amount of defective product in addition to whatever amount of defective product results from special cause variation. Let us examine how this information is used in practice in several situations.

Capable Processes

When the full range of the natural output from the process as predicted by the statistical model is contained within the range of the product specification, the process is a capable process for producing that product. Absent special cause variation, it is expected that all of the output from the process will meet the product specification. When your plant enjoys this situation, you can anticipate that your only source of defective product will be special cause variation. In this case, you have four different paths for improvement:

1. *You can elect to improve the inherent performance of the process further in order to do new things.* In that way, you can enable a shift in the product specification such as was done in the vinyl sheet example. Alternatively, the improved capability will enable you to produce new products with more stringent requirements as we did at Exxon Baytown. Your process model will enable you to conduct designed experiments that will identify the sources of inherent variation within your process. Resolving that variation allows you to improve the process so that your plant can work in new ways.
2. *You can elect to improve the inherent performance of the process in order to make it more robust.* Processes that are capable, but not

very capable, frequently produce defective products that result from special causes. By making your process still more capable, you can provide some added tolerance between normal process performance and product requirements. In that way, you can often enable your operating team to resolve special cause events *before* they have an adverse impact on the product or the customer.

3. *You can improve your performance by eliminating recurrent special causes.* Many capable processes routinely produce defective products as the result of frequently occurring nonsystemic causes. When you know that your process is inherently capable, you can focus your improvement efforts on identifying and removing special causes.
4. *You can redeploy your improvement efforts to another process.* Many plants have some processes that are capable and others that are not. A capable process in one part of your plant allows you to focus your engineering and other special resources on operations that have greater need and therefore will produce greater return from your improvement efforts. Meanwhile, sustained operation and routine improvement of this capable process can continue as an activity of the frontline teams.

In addition to enabling an analytically structured approach to improvement, using the information provided by the statistical analysis allows you to make an informed selection of the improvement actions that will have the greatest benefit to your business. Moreover, depending upon the improvement path you choose, you can also intelligently determine the urgency and the resources that will be assigned to that path.

Incapable Processes

Alternatively, it is common to find that the natural range of output from your process exceeds the range of the specification for one or more of your products. In this situation, the process is inherently incapable of routinely producing that product without defects. Because special causes of variation also routinely arise in manufacturing, you can anticipate that, when you have an inherently incapable process, you will produce defective products that result from both special cause variation and systemic variation.

Faced with a process that is inherently incapable of producing products that routinely meet your customer expectations, the process model will once again give you the information you need to select the improvement method most appropriate to your business needs. The list in this situation is generally the same as the list for capable processes. You can elect to improve the process to make it capable of doing new things or to make it more robust. You can elect to remove special causes by deploying your engineers or by relying on your frontline teams. You can make an informed selection of the most valuable focus for your improvement efforts.

The key difference between selecting a path forward for capable processes and incapable processes is the recognition that, with an incapable process, you cannot routinely produce all good products until you improve the production system. Simply focusing your frontline teams on careful operation combined with the identification and resolution of special causes will not produce products that consistently meet customer expectations.

Unless incapable processes are a common problem in your operation or you are willing to disappoint your customers some of the time, your engineers and managers urgently need to focus on improving incapable processes. As the professionals make systemic improvements, the process model will guide, assess, and validate those improvements. Even with an incapable process, your operating teams can use the process model to identify and resolve the sources of special cause variation as they occur, regardless of whether you elect to deploy your engineers to process improvement. That new action at the frontline will initiate some valuable and immediate improvement even without engineering involvement.

However, it is important to recognize the need for engineering or management participation in the improvement of incapable processes. Frontline teams can help by resolving episodic or recurrent special cause variation, but incapable processes in the chemical industry are rarely resolved by frontline action alone. Fortunately, if you have some capable processes, you can redeploy your technical assets to the problematic processes with no detriment to capable processes that exist elsewhere. Understanding the capability of all your processes allows you to balance your resources intelligently across your entire operation in a way that produces the best business result possible with the resources available.

Key idea: Many process plants that do not have a meaningful ability to assess the inherent capability of their processes have a strong tendency to treat all parts of the plant as if they are equal in value or importance—or at least nearly equal. This means that all areas of the plant will have a fair share of the available technical support. The difficulty with this scenario is that most businesses generally achieve far better results by focusing technical support on the areas of greatest importance if they can be identified. Statistical process control modeling of your processes provides the capability to do that with some confidence that you are making the right choices in a complex situation.

Case Study: Reliability Engineering

At Suncor, at the beginning of 2008, we practiced a very traditional approach to reliability engineering. Our entire plant exists for the unified purpose of producing material that flows through several process areas to become essentially a single finished product. Therefore, each of the major operations had very similar production goals and very similar asset value. Therefore, we assigned very similar engineering teams to support each area of the plant.

Once we began statistical assessment of performance, we found that, by a very wide margin, our most valuable improvement opportunities were centered in the upgraders that transform bitumen to synthetic crude oil. With that information, we promptly and substantially reallocated engineers toward the upgraders. Because of that shift in resources, within the first 6 months we achieved a great improvement in plantwide production. During the second 6 months following that redeployment, we achieved a new record for production. Statistical analysis of our operations enabled us to deploy our companywide assets in a focused way that produced the best possible outcome.

Many statistical methods are of value to the practice of SPC, including the designed experiments that enable the identification of the few most important sources of systemic variation from among the many possible sources of variation. In addition, a rich portfolio of technology and programs has been developed around the theme of six sigma. However, the process capability model is the essential concept that leads us to understand the value proposition of SPC. Consistent with *lean enterprise thinking*, as you and your team begin to practice *statistical thinking*, you are certain to experience many improvements that do not require statistical analysis.

Case Study: Statistical Thinking

In one effort at Exxon Baytown, we intended to improve the performance of our critical process control instruments. We defined a critical instrument as one that affected safety or quality. As part of this activity, we reviewed the historical records of routine maintenance for those critical instruments. During that review, one of our technicians recognized a pattern that did not require statistical analysis, but only statistical thinking or statistical understanding.

The records showed that, in virtually every case where a technician performing routine maintenance adjusted the sensitivity of an instrument by a half percent, within either of the next two routine maintenance visits, the technician adjusted the sensitivity in the opposite direction by the same half percent. As our technician recognized, that is a classical symptom of adjusting a stable system although it is performing as expected; the principal source of variation in the system was unnecessary operator adjustment.

Essentially, we learned on further investigation that for high-quality electronic instruments, the inherent capability of the instrument is such that constantly recentering the range of the instrument has no value. When high-quality electronic instruments fail, they do not fail by drifting a half percent away from their set point. They fail by stopping completely or by providing clearly improper values. This understanding of our data enabled us to change the nature of our routine maintenance of critical instruments from a program of constant adjustment to a program that spent more time on the care of the instrument. When we stopped adjusting perfectly good instruments and began to care for the environment, installation, and life cycle of the instruments, we got much better performance with much less effort. In this manner, we had a very successful program of improvement and we were able to expand that program to include more instruments than originally were within the scope of our effort.

USING SPC AT THE FRONTLINE IN A PROCESS PLANT

Statistical process control coincides with the lean values concept that the tools of improvement should enable everyone's participation. Engineers and managers can improve the inherent capabilities of the process and frontline teams can identify and remove special cause variations as they occur in operation. In the process industries, improvement derived from the efforts of managers and engineers is a basic capability. Therefore, it is likely that the most important new contribution of both lean and SPC is to enable autonomous frontline teams to contribute to the improvement process. This is the capability that will distinguish you from your competitors and SPC does this in a number of ways.

Assuming that someone, probably a statistician or six-sigma black belt, has provided the operating team with a mathematical model of its process, the frontline team can now do many things it could not previously do. Normally, the mathematical model is deployed to the frontline in the form of a tool called a "run chart." In many chemical and process plants, this statistical model is configured within the unit console as control algorithms. Whether process technicians perform the analysis or it is done for them as part of the console programming, it is useful for you and your team to understand the nature of the exercise.

Key idea: The run chart continuously refreshes the statistical thinking and understanding your team has about how its process should operate and how it is actually operating in comparison to expectations.

Using a Run Chart

The run chart is nothing more than a continuous, *real-time comparison* of the *expected performance* of the operation *as predicted by* the statistical model *with the actual performance* achieved as the unit operates. That comparison is supplemented by a statistical understanding of what the various comparisons mean. The practical element of SPC practice at the frontline is a series of "run rules" that describe the proper action to be taken in response to each operating scenario described by the run chart information. This analysis provides frontline teams with new understanding of the real-time performance of their operations and new responses to drive improvement. Let us see how this works in practice.

When the Run Chart Says the Process Is Operating Normally

The process is determined to be operating normally when the performance achieved is in accord with the expectations of normal performance as described by the statistical model of the process. Whether the production process is a capable process for the current product or an incapable process, in normal operation, it should produce results within the range of values predicted by the process model and randomly distributed within that range. That is how the process model was constructed. As a result, when the operating team collects current performance data that exhibit the expected range and distribution, its members will know that the process is operating normally. This knowledge is very valuable in determining the actions of the operating team as prescribed by the run rules.

Run rule: When real-time data demonstrate that the process is operating normally, there is nothing that the operators can or should do to change process performance. Operators should devote all of their improvement efforts to the care of the equipment (see Chapter 9) or other routine activities.

For operators of *capable processes,* this run rule makes a lot of sense. Everything is fine. Leave it alone. For capable processes that are running as expected, any adjustment is more likely to do harm than good and the operators can easily understand that.

This run rule is more difficult for operators of *incapable processes* to understand or accept, but it applies equally to them. If an incapable process is producing results that are all randomly distributed within the predicted performance range of the process, it is *natural and expected* for an occasional performance result to fall outside the customer specification. When this happens, most operators want to adjust the process; however, in this case, the statistical understanding remains that such an adjustment is likely to make ongoing performance worse rather than better by introducing the special cause impact of altering the center of the distribution.

For example, if the real-time production sample shows a result that is out of the product specification range to the high side of the range, it is normal for the operators to want to compensate by adjusting the center point of the process toward the low side. Unfortunately, if the process is performing normally, it was originally *centered* within the specification range of the product. Adjusting the center point toward the low side will indeed reduce the number of results that are normally out of range on the high side. However, *it will increase the production of product that is out of range on the low side by an even greater amount because a special cause bias toward low-side production has been created when the process was recentered.*

Your frontline teams can make other valuable contributions to improve performance when an incapable process is operating normally. Recognizing that they have that opportunity and further understanding that they should not attempt to produce performance that exceeds the capability of their process allow them to capture the value of those other contributions. Adjusting an incapable process that is operating normally has no value and perhaps makes performance worse.

Key idea: Whether the process is capable or incapable, adjusting the process when it is operating as it is expected to operate will normally make the situation worse, not better. You and your operators need to understand this or you will be chasing performance that you cannot achieve, perhaps making things worse, and wasting efforts that could be better spent on other things.

When the Run Chart Says the Process Is Producing an Unexpected Result

The second opportunity to use the statistical comparison of real-time results to the process model is when the process produces an unexpected result—that is, when routine monitoring of current process performance produces a result that *is not* predicted by the process model. Again, whether the process is inherently capable or incapable, the run rule that drives frontline actions is the same. Whenever the process produces a result that is not within the range of predicted results or not properly distributed within that range, the operators can know with certainty that a special cause external to the process is producing that result.

Run rule: Whenever the real-time process result indicates performance that is not predicted by the process model, the operating team knows that a special cause is affecting performance. Therefore, team members must identify and eliminate it as rapidly as possible so that the process can return to normal operation.

When the process is operating normally, the frontline team will make the situation worse if it intervenes; however, when there is a special event, it typically *must* intervene. Processes affected by a special cause generally require some immediate action by the frontline team in order to return to normal. Even if the special cause is transient and the process will recover by itself as the event passes, *prompt action* by the team *to isolate and identify the special cause while it exists* is a necessary proactive step if the team is *to prevent that same event from recurring* in the future. Some processes experience frequently recurring special events. Resolving a recurring event or preventing an episodic event from becoming a recurring event will have an operating impact similar to inherent improvement of the process.

The teams are assisted in their search for improvement of special causes of variation because they know that whatever caused the abnormal performance is not common to their process; therefore, they are looking for something that has changed. Further, if the team continuously and carefully monitors performance, it also knows that whatever is causing the abnormal performance is happening at that time. That close correlation between cause and effect is a great benefit to problem identification and resolution. Many teams detect changes in performance so quickly that the life span of an abnormal event before it is identified and resolved is quite short. By reducing the frequency, extent, or duration of abnormal events, teams can again have an impact that is similar to improving the inherent process.

Frontline teams are uniquely able to contribute to this improvement opportunity. Only frontline teams are always present and only they can always correctly answer the questions "What just changed?" or "What new is happening now?" By providing your frontline teams with a model of expected performance that enables them to transform routine process observations into a *real-time search* for abnormal variation when it exists, you will create a truly powerful improvement capability that you cannot access in any other way.

When the Run Chart Says the Process Is "Nearly Normal," but Results Are Drifting

The third opportunity for benefiting from the statistical model and statistical thinking to help your frontline teams is when the real-time process results are contained within the range of predicted performance, but the results are no longer randomly distributed within that range. Again, this situation and the appropriate responses are common to both capable and incapable processes. Two classic examples are generally used to illustrate this situation:

1. *All of the real-time results are within the predicted range, but all of the most recent results are on the same side of the normal middle of the range.* This result usually indicates that the process needs to be recentered. Although it is inappropriate for the operating team to recenter a unit that is performing normally, when team members detect that performance is abnormal—specifically, when they detect that the center of the range has drifted away from the intended target value—*adjustment* is not only entirely appropriate but also *mandatory.*

2. *All of the real-time results are within the predicted performance range*—they are even fairly distributed on both the high and low sides of the expected center of the distribution—*but there is a clear nonrandom pattern to the most recent results.* This nonrandom pattern is likely to appear as a trend in the results; for example, several recent points may be fairly distributed on both sides of the mean but, considered together, they are demonstrating consistently increasing values. Because the statistical model of expected performance predicts that the performance results will be randomly distributed within the range, if the results demonstrate a clear trend within the range, then something is not normal. Unlike the situation where all of the results are suddenly on the same side of the center point, the nature of what has happened is not intuitively apparent, but clearly something has happened and the team needs to intervene to identify the source of the variation and resolve it.

Either of these two situations is only observable through a subtle analysis of performance that certainly would be impossible at the frontline without statistical capabilities to assist the team. With this analysis, team members can intervene to return the process to normal operations, even when no individual measured result has yet demonstrated that the process is not performing normally. By enabling a proactive response to correct an operating problem before any adverse result has occurred, SPC provides your operating teams with a great new capability. In a situation like this, no independent tangible event has yet occurred to indicate the need for a reaction, so no other improvement effort could ever match this result.

Run rule: When the center of the range of production changes or when the results within the range of current production become nonrandom or demonstrate a clear trend within the range, some "special cause" event is happening that requires the team to respond. The response may be a simple recentering or something more complex. The operating team needs to identify and resolve the cause of the abnormal results.

Through careful analysis and monitoring, the team has an opportunity to respond *before* any production consequences occur. Statistical process control provides your frontline teams with the ability to recognize and distinguish situations where the process is operating normally and adjustment is improper from those where a special cause is affecting production and adjustment is mandatory. Again, this is a proactive capability that cannot possibly be created within a traditional reactive process management regime.

AVOID THE WASTE OF EXCESS QUALITY

Lean philosophy is often described as driving waste and excess from manufacturing. One surprising waste that is becoming more common is the waste of excess quality. Statistical methods are a powerful and even enjoyable capability for both engineers and operators. In some situations, the tool is so attractive and powerful that some enthusiastic practitioners adopt product and process improvement almost as if it were an independent goal separate from the objectives of the business. It is important to remember that *the purpose of improving process capability is to make the business better.* Once the process is fully capable and meets customer expectations with a high degree of robust performance, you and your teams should consider the possibility that the next increment of improvement should be devoted to making a different process more capable rather than making further improvements to an already capable process.

Alternatively, you may want to find another type of improvement altogether, such as improving production capacity or introducing a new product. Your frontline teams will certainly continue routine process monitoring and routine actions to search for and eliminate special causes. However, you may want to redeploy your engineers and you may want to redirect the improvement focus of your frontline teams.

Case Study: A Cautionary Tale—The Price of Perfection

The classic tale of caution in this arena is the Motorola Bandit pager. During the period when Motorola was almost religiously devoted to six-sigma methods, the Bandit product and the production process for it were so thoroughly developed and so very robust that Motorola was able to market it as a product that would have a mean time between failures in excess of 150 years. Unfortunately, the Bandit pager was a classic radio pager; while

Motorola engineers were making it nearly perfect, the competitors at Nokia introduced the digital cellular pager. It turns out that no one really wanted a pager that could last 150 years and the effort to produce a perfect radio pager distracted the development team from recognizing the new digital technology that would soon dominate the market.

This description of statistical methods was brief, and you probably noticed that I did not include any math or even any charts and graphs. That was deliberate. My goal was to introduce you to statistical thinking. Statistical process control is a powerful tool and statistical thinking will be very useful to your engineers as well as your frontline teams as a companion to your lean implementation efforts.[1]

NOTES

1. If you want the details of statistical practice, many books on the topic are available. One of my favorites is *Six Sigma: The Breakthrough Management Strategy*, by Mikel Harry and Richard Schroeder (New York: Broadway Business, 2006).

8

Mistake Proofing or Poka-Yoke

INTRODUCTION

The technology of mistake proofing, known as *poka-yoke* in Japan, where the term was coined, is very closely associated with lean manufacturing because of its great ability to enhance the improvement efforts of frontline teams in simple and effective ways. Improvement at the frontline is a quintessential lean value.

Mistake proofing is thoroughly documented in Shingo's book, *Zero Quality Control: Source Inspection and the* Poka-Yoke *System*. Similarly to his SMED (single minute exchange of dies) book, Shingo describes *poka-yoke* in both theory and practice, and he provides many specific examples. In most situations, you can find an example in Shingo's book that is sufficiently close to the problem that you are trying to resolve that you can simply select an existing solution rather than create a new one. In fact, *poka-yoke* is one of those techniques that is so easy to understand that it can be introduced with only a brief theoretical discussion and a few examples to illustrate the practice.

Much like SMED, there is very little in the concept of *poka-yoke* that most people have not seen before or even practiced in limited circumstances. The breakthrough in understanding that makes mistake proofing such a powerful tool of improvement is that it formalizes and documents the practice as an easily understood theory supported by a variety of examples appropriate to many different situations. *Poka-yoke* enables many people to use a capability frequently and routinely that only a few people used episodically before.

Because it is so straightforward, *poka-yoke* is often among the earliest skills deployed to frontline teams at the start of autonomous improvement. They can use it immediately with great value while both they and you can be certain that they will always be doing the right thing in the right way: *Poka-yoke* does not change the value stream of the work or the work product; rather, it simply ensures far less chance of making a detrimental mistake.

As with other technologies of lean practice, this tool is especially valuable in the process industries. In most mechanical manufacturing, many people are tied to a paced line for assembly or other work that requires a fixed rate of precisely predetermined effort. In our process plants, most people have assignments that are less specific. Those assignments provide a freedom of action that allows them to create and use slightly new methods of working such as this in order to ensure that they produce a perfect outcome. In addition, we have fewer truly routine work elements than mechanical manufacturers do, so providing our people with a capability to ensure that nonroutine work is routinely done well has special value to us.

MISTAKES COME IN TWO PARTS

The essential concept underlying mistake proofing is that mistakes come in two parts. The first part is when a mistake occurs and something in your plant or process is not as it should be. The second part is when the situation created by the mistake is allowed to mature until a meaningfully detrimental consequence results. This understanding of the mechanism of mistakes gives us two opportunities to avoid the bad consequence. The first is not to make the mistake in the first instance and the second is to recognize that a mistake has occurred and to correct it before it matures into a bad consequence.

Often the ability to intervene successfully in the mistake process is substantially different in each of these cases. Recognizing that they both exist is important because it allows us to make a purposeful selection of the better opportunity between the two or even to intervene in both opportunities in the event of a potential mistake that we especially want to avoid.

Note: As indicated, this technology was developed to avoid mistakes, but it is applicable whenever "something in your plant is not as it should be."

Whether that situation exists as the result of a mistake or for some other reason, *poka-yoke* practices allow your technicians to recognize the situation and intervene to avoid the consequences. (This is discussed further in Chapter 9.)

THE CONSEQUENCES OF MISTAKES

Some mistakes are essentially meaningless. When we were first married, my wife practiced law using her maiden name of Wolowic, so she adopted the practice of introducing me using my full name: "This is my husband, Ray Floyd." As a result, many of her professional colleagues who knew her and not me assumed that my last name was also Wolowic and that I had one of those Southern two-part names like Jim-Bob. Because I saw her colleagues infrequently, for years they would greet me with something like, "Good morning, Ray-Floyd." Mistakes of that sort are often amusing but generally have no consequences and do not require any resolution.

Other mistakes do carry consequences. In the process industries, those consequences can range from inefficiency to poor product quality to fire and explosion, and perhaps even loss of life. Mistakes with significant consequences merit formal intervention to prevent those possible consequences from becoming reality. Providing your team with a good way to avoid the mistake or the consequence that could otherwise arise from the mistake can have real value to your operations.

Key idea: The industrial importance of mistakes is measured not by the fact that a mistake is possible or even likely, but rather by the potential consequences of the mistake if it occurs. When we are managing mistakes, our real concern is managing the *consequences* of mistakes rather than the mistakes themselves.

As you deploy this powerful and enjoyable new capability, it is valuable to formalize the understanding that you are not interested in creating an anal-retentive group who worries endlessly about every possible mistake. You are interested in creating a team that can avoid mistakes that have business consequences.

Mistake Proofing: Preventing Consequences

The new opportunity that *poka-yoke* presents is to intervene between the occurrence of a mistake or event and the evolution of the mistake into a bad outcome. In that way, the mistake has no consequences and provides a learning opportunity. As the practice of mistake proofing matures, you will find that people who repeatedly experience an intervention immediately after a mistake soon begin to identify and correct future mistakes before they occur. As I write this, my computer constantly tells me when I misspell a word by inserting a red line under it. This allows me to correct it immediately. Repeated intervention of this sort has taught me to spell more words correctly on the first try. As I experience this form of real-time mistake proofing, I learn not to make mistakes. The same thing will happen in your plant.

Mistake Proofing Is Common Knowledge

Separating mistakes into two parts and using the new opportunity presented by that separation to prevent a consequence is a common experience that your people already know well; this makes it easier to introduce it successfully into your plant. Once you explain it, you will find that all your people already know what mistake proofing is and how to use it. All that you need to do is to teach people to recognize existing industrial opportunities to apply this concept and to create new opportunities for themselves where they do not exist today. Although most people already know all that they need to know in order to practice *poka-yoke,* most simply never think to do so at work.

The concept is simple. My wife and I learned from our mothers (and I suspect that you did as well) that, when you add an egg to a mixture, you break the egg into a separate bowl and examine it before adding it to the mix to avoid the consequence of a bad egg or a piece of broken eggshell spoiling the mix. This practice is common because everyone knows that it is possible to come across a bad egg and, if a bad egg gets into the mix, the whole thing is ruined. They also know that it is easy to stop that from happening. In mistake-proofing terms, this use of an intermediate bowl is described as "physical separation." It is one of several standard techniques used to prevent a mistake—in this case using a bad egg—from maturing into the bad consequence of ruining the mix. My mother and many others were practicing *poka-yoke* before Shingo was born!

There are also many well-known commercial examples of *poka-yoke* that everyone recognizes and understands. Most people are familiar with the events of autumn 1982 when Tylenol packages were poisoned and several people died. That series of events gave rise to the *tamper-evident* packaging that we find today on most foods and drugs. As with the eggs, tamper-evident packaging is an example of separating events into two parts. In this case, rather than physical separation, the mistake-proofing concept is to provide a "visual signal" that the situation is not as it should be. Tamper-evident packaging is a good example that this technology is useful not just for preventing mistakes but also for enabling your operators to recognize any situation in your plant that is different in some way from what it should be. That capability often has immense value in the process industries.

Tamper-evident packaging is also a good example of the difference in difficulty of actions that makes events apparent in a way that *avoids* consequences *after* a mistake and the difficulty of attempting to make the mistakes or events *impossible. Tamper-proof* packaging that would make it physically impossible to interfere with the product is extremely difficult and expensive. A friend and former director of product development at Procter & Gamble advises that they worked on this problem for a number of years and finally gave it up as impractical.

On the other hand, *tamper-evident* packaging is often no more difficult or expensive than incorporating a visually apparent plastic band that ensures buyers that the package has not been previously opened. The tamper-evident system eliminates the need to take the difficult and expensive steps required physically to prevent someone from tampering with the package. As long as it is always evident that a package has been opened, the consumer will reject it in favor of a sealed package and no harm will occur. Because they recognized that there are two separate opportunities to prevent harm, food and drug manufacturers have been able to take the easy and inexpensive route with complete success.

WARNING SYSTEMS

Shingo's book describes two essential types of mistake proofing: *prevention systems* and *warning systems.* However, we will describe only the use of warning systems intended to make it apparent that a mistake or event has occurred or is about to occur. In this way, they allow the operators a second chance to intervene in the progression of a mistake into a consequence in a way that will avoid the consequence. Prevention systems are very similar in concept *except* that they are largely automated and, when an anomalous situation is detected, the system automatically stops production until the anomaly is corrected. Essentially, the prevention systems are an automated form of the lean practice of *andon*. For all the reasons described in Chapter 1, systems that stop production are often completely inappropriate for use in the liquid industries, where stability and continuity of the process are typically far more important than they are in mechanical manufacturing.

The value of warning systems for mistake proofing is that they are quite effective and generally so simple that anyone can adopt them. Further, *poka-yoke* used in this way does not change the essential character of the work or of the work product. The only change is the introduction of a signal to alert the operator that the immediate situation is not precisely as it should be. Because mistake proofing does not change the work and does not have an impact on the outcome of the work except to make it more likely that the outcome can be perfect, it is a valuable tool for early use in autonomous improvement. You can give your teams a new capability that most people will enjoy using creatively and that is already completely within their existing skills and authority. As they use this new capability, both they and you can be confident that, without significant management intervention or oversight, this new practice will produce only good outcomes.

Four Types of Warning Systems

The four essential types of warning systems are physical separation, visual signals, pattern recognition, and simple devices. Each will be described in both theory and example. The examples used in this chapter all come from process plants so that it is clear that you can use this technology in your plant. However, most of the examples from mechanical manufacturing contained in Shingo's book can also be adapted for use in process plants.

Poka-Yoke *Practice 1: Physical Separation*

Similar to the egg in the intermediate bowl, there are many easy ways physically to separate mistakes from consequences. A well-known example in the liquid industries is the process safety management (PSM) practice of identifying the occupied structures within the battery limits of an operating unit and assessing the need for the people in those structures to be physically present. According to PSM, people who do not need to be in hazardous areas to do their work should be relocated.

In the case of PSM, possible mistakes or events include gas release, explosion, fire, or other catastrophes that might occur in the plant. Certainly, every responsible operator takes all possible care to prevent such an occurrence, but the PSM standards go further. They recognize that even with due diligence, it is unlikely that all events can be perfectly avoided in all plants all the time. Because PSM standards require that process plant owners move people who are not operationally essential away from the unit, they physically separate those people from the possibility of injury arising as a result of imperfect operation. Should an event occur, people who are not physically present cannot be injured.

Key idea: This practice does not conflict in any way with the process industry philosophy that all injuries can be avoided. Placing people where they cannot be injured is a perfectly legitimate way to avoid injury.

Case History: Physical Separation—Large Events

Most of the people killed during the explosion at BP's Texas City refinery in March 2005 were in a meeting that was not essential to the operation but was nevertheless held within the battery limits. There are still unresolved issues surrounding the explosion, but the fact is that those people did not need to be there at that time, and if they had been elsewhere, they would not have been hurt. Physically separating the possible consequence (injury) from the possible event (loss of containment) would have been a perfect way to avoid this catastrophic outcome.

Fortunately, not all examples of physical separation are so dramatic. Most are small indeed, and it is in the small event arena where your frontline operators can make contributions that aggregate to large operational improvements. Once again, the ability of *poka-yoke* to allow people at the frontline to manage the conduct of your business to produce perfect results at a level of detail that could never be achieved by your engineers and managers is a classic example of the value of the lean effort.

Case History: Physical Separation—Small Events

As part of Exxon's polypropylene production process, we incorporated a wide variety of additives during the last production step. Because the additives were often unique in some way, we normally compounded our own. As we began to adopt very precise expectations for our product quality, we noticed that, although our processes for polymer production were highly technical and complex, many of our product quality problems occurred in the relatively unsophisticated step of compounding the additives.

In one interesting case, we found that the compounding recipe included measuring 7 pounds of an ingredient into each batch of additive. The operator performed this step by using a scoop by hand that had been "calibrated" as a 2-pound scoop. We immediately assumed that the measurement problem resulted from the inability of the operator to estimate the seventh pound in the recipe adequately, which required approximating half of the 2-pound scoop. We began using a 1-pound scoop. Imagine our surprise when we found that the smaller scoop made the measurement problem worse, rather than better.

We then discovered the operator had been adequately estimating a half scoop within the tolerance required. He was not equally good at counting. Rather than missing the recipe by part of a half scoop, the operator periodically missed by a whole scoop in either direction (more or less) because he had miscounted the scoops as he added them to the batch. Switching to the 1-pound scoop increased the frequency of error because it increased the likelihood of a miscount.

As the compounding process was originally designed, the operator scooped from the bulk supply of raw material directly into the compounding mixer. Whenever he lost count, he had no way to recover. In *poka-yoke* terms, the mistake of miscounting was physically linked to the consequence of adding the wrong amount of material to the batch. In the event of a mistake, the operator did not have a second opportunity to prevent the consequence.

The solution was to separate the mistake physically from the consequence. In this case, we considered placing a scale in the compounding area specifically to measure the full amount of the material in aggregate before it was added to the mixer, but that would have greatly increased the amount of operator travel (walking to and from the scale) and that degree of precision had no real value to the process. We found that we did not require true precision; we just wanted to avoid the gross imprecision of miscounting the scoops. The operator adopted something even easier and more flexible. He acquired a small plastic bucket that he marked at various levels. He was able to carry his bucket around among his several bulk containers and use it for many different materials and recipes. From that point forward, he scooped material from the bulk supply into his bucket before he added it to the compound.

With the help of the markings on his bucket, he was able to get a visual representation of the measured amount in aggregate and that provided him with a second chance to get it right before he placed the material into the compound. In the event that he was unsure of the amount, he simply poured the material back into the bulk container and started counting again. In that inexpensive and easy way, he created a physical separation and a second chance to ensure that he always added the correct amount. That practice seems a lot like breaking eggs into a bowl, does it not?

Key idea: Our operator was clearly more than able to count to seven. The problem was not one that could have been solved by training or even by close supervision. As we all experience from time to time, the operator was distracted by another person or event or simply did not pay perfect attention to his work. By using the bucket to separate the mistake physically from the outcome, this imperfect individual became capable of producing perfect work.

One way that business leaders routinely fool themselves into poor performance is by creating jobs or systems that are far more intense than people can reasonably perform perfectly on a routine basis. A task that has several closely linked critical steps is a good example of a task that is at high risk. As you deploy the methods of *poka-yoke* with the expectation that work can be perfectly executed, people performing such intense assignments will get another chance to deliver perfect performance. By allowing them to lower the intensity of their work modestly through creation of a physical separation such as that described previously, *poka-yoke* will make a real difference for the people and for the business. This freedom for people to change their work to make the outcome perfect is a real advantage that process industries enjoy when compared to the paced lines of mechanical assembly.

Poka-Yoke *Practice 2: Visual Signals*

Another excellent technique for mistake proofing is to provide visual signals that display information about the state of the plant. These signals do not have to be highly sophisticated devices or otherwise difficult to conceive or implement. They only need to provide useful information that allows operators to prevent mistakes or to identify and recover from a mistake before it evolves into a consequence. This practice has great value beyond resolving mistakes because it can often be deployed to identify other anomalous conditions in your plant that do not arise from mistakes, but nonetheless require recognition and response.

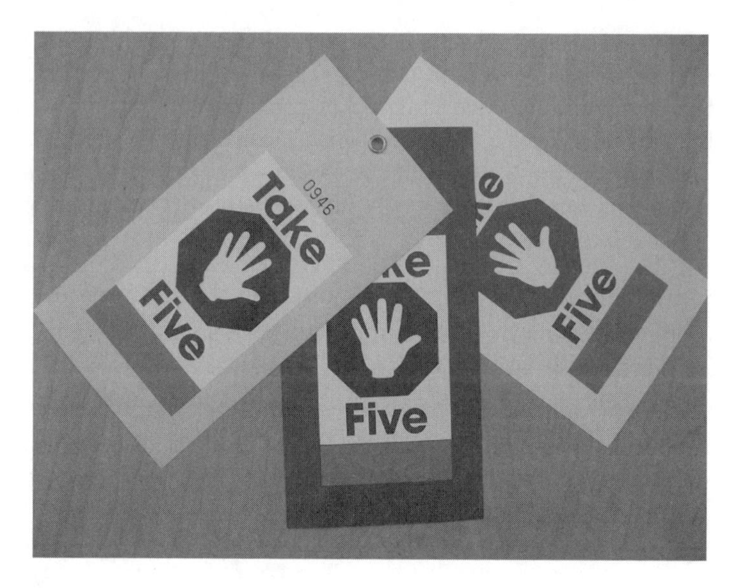

FIGURE 8.1
Multicolored tags used to ensure accurate fieldwork.

Case History: Visual Signals to Ensure Proper Maintenance

One application of this method that we have adopted at Suncor ensures that when we open our hydrocarbon lines for service, we always open the correct lines, do the right service, and secure the lines properly before returning them to service. We use a multicolor tag that is physically attached to each flange to be opened. Figure 8.1 is a sample of such a tag (the four colors appear here as shades of gray). For illustration, let me describe the process of opening a line to replace a valve:

1. The planner initiates the process by giving the full four-part tag to the job walker when the flange is ready to be opened. Each tag has a control number, which is recorded by the planner in the job instructions provided to the technicians who will do the work.
2. The maintenance job walker, along with a process technician, locates the flanges to be worked on and agrees with the process technician that they are the correct flanges and that they are ready for work. At that point, the job walker and the process technician attach a tag to each flange. Each part of the tag is a different color. In this case, a yellow tag indicates that the flange is ready to be opened.
3. The mechanical technicians then can identify the flanges to be serviced by observing a yellow tag and comparing the tag's control number to the number in their job instructions. If they match, the technicians open the flanges, remove and replace the defective

valve, close the flanges, and remove the yellow tag. When the work is mechanically complete, they remove the yellow tag and thereafter a red tag will be visible on each flange.

4. When the flanges have been mechanically closed, a process technician, along with a mechanical quality control inspector, examines the work to ensure that it is ready to return to service. When the flange has been successfully inspected, the red copy of the tag is removed and a white tag is visible.
5. When the unit has been completely inspected, we begin to circulate hydrocarbon and the process technicians again visit each flange with a white tag to ensure that the flanges are secure in operation. At this point they remove the tag. The unit then displays no visible signals that work is required.

In this way, we reduce the possibility of confusion. By visual signal, we can confirm that all flanges have been approved for work, have been completed and inspected in their turn, and, finally, that the whole unit is uniformly ready to return to operation. By returning the tags as they are removed from each flange to the planner, who then matches them with the serial number that he originally issued for the work, we have a third party who can track the job as it progresses and confirm that all the work at each step has been completed.

Thus, we have made it visually apparent at each point that all valves and flanges have received the appropriate attention required to progress to the next step. Moreover, a visual signal that any one of the many flanges has not received the full program of work provides a second chance to get it right *before* we proceed to the next step.

Although this system does not physically guarantee no mistakes, the presence of colored tags on the unit gives us a clear second and even third chance to identify where mistakes might have been made and correct them before the work proceeds. As a result, it is highly unlikely that the wrong flange will be opened or that an incomplete or untested flange will result in a leak when the unit is restarted. That is far more assurance than we have had before. Since adopting this simple practice, we have completed more than 4 million hours of mechanical work, including several major outages, without opening the wrong flange or returning a leaking flange to service.

Color and Shape as Visual Signals

Color and shape are two great variables that allow people to recognize change or progress in the state of the operation. In blending additives, Exxon now often takes the extra step of using pellets of different sizes or a mix of pellets, granules, crystals, and powders to differentiate various materials. This gives the operators greater certainty and a second chance to make the results perfect by inspecting a mix to ensure that it contains all the components it should contain.

Before intentionally creating this visual difference among the components, we previously had a compounding practice that mixed several white powders together; once the mix was complete, there was no possibility beyond chemical analysis to ensure that it was correct. By using distinctive textures or shapes, the state of the mix is instantly apparent to anyone who wants to look. It sometimes costs a bit more to do this, but often it does not. The materials would always have some texture. We simply wanted to choose those attributes in a way that helped us avoid mistakes. The certainty that the work can be done perfectly generally makes up for any small extra cost.

Color coding things to demonstrate visibly that they go together or that they are different is a wonderful way to bring more certainty to any process. It is often possible to differentiate or match ingredients by color and obtain the same effect as you would by changing the size or shape of the material. Where physically differentiated ingredients are not practical, we can achieve a similar improvement by color coding the containers. In one or more of these ways, we give the blending technicians an opportunity for visual inspection that lets them know that they are compounding the right ingredients. In the situation where the colors, shapes, or textures survive into the compound, they can visually examine the finished additive to gain further assurance that they have compounded the correct materials. This practice often allows the operating technicians, who receive the additive from the blending technicians, to look at it and ensure that it is what they expected to receive. This is yet another chance to avoid a mistake.

Case Study: Color Coding to Facilitate Autonomous Lubrication

At Exxon, we follow a practice of owner-operator lubrication. Color coding is an important part of that effort. Although in large plants, we naturally have a wide variety of lubricants for different applications, by color coding the entire system of containers and transfers that a lubricant follows throughout the plant, we can be sure at each step that we are using the right lubricant in the proper application. For example, gearbox oil for pump drivers comes out of a blue drum in the shop, placed into a blue field dispenser, and, ultimately, put into a blue opening on the gearbox.

If, at any time, the colors on both the dispenser and the receiver in any transfer of lubricant do not match, then the operator's authority to make the transfer fails. In that situation, the operator has a clear indication that there is a problem and the lubrication transfer does not proceed until the problem is corrected. In this way, we facilitated a high degree of operator care maintenance and have a greater degree of assurance that we have a perfect outcome than we had previously when a professional "oiler" traveled around the plant performing the task without these visual signals.

Once they are familiar with this practice, your teams at the frontline will develop many great ways to make their work visually apparent; the result will be that they can give themselves a second chance to make the work perfect all the time. Visual indicators allow people to identify and to correct errors promptly, which make perfection much more likely.

Case Study: Prescription Medicine

Using size, shape, and color to differentiate one thing from another is a form of *poka-yoke* with which we all have personal experience. When we purchase a prescription drug, it always has a distinctive appearance, and a representation of the proper visual appearance is a part of the drug advisory delivered to patients along with the medicine to provide a basis for patient recognition and confirmation. Because of the distinctive appearances and markings on the drugs, pharmacists and even patients have a second chance to assure that they are receiving and taking the proper medicine. When I introduce *poka-yoke* to people, I constantly remind them that they are adopting a practice that they already understand.

Poka-Yoke *Practice 3: Pattern Recognition*

Most people can recognize complex patterns and respond to the signal presented by those patterns. The secret to success is to teach people to recognize patterns that exist naturally or to allow them to create meaningful new patterns that are especially easy to recognize. Often management can initiate this practice by establishing a few patterns that can be exemplary or ubiquitous because they are broadly applicable to many of your operations.

As described previously, colors and shapes can assist the compounding technician to prepare the correct material perfectly. However, that practice is also an example of using pattern recognition to ensure operating technicians that they have received the proper material. After the additive compound is delivered to the operating team, the team member responsible for loading it into the feeder can take a handful for a quick visual inspection. If the proper combination of colors, shapes, or other visual signals is apparent, that pattern tells him or her that this is likely to be the right material, prepared in the right way. Many operators develop sufficient recognition of the visual pattern of the material that they can recognize when the proper ingredients are present in the proper proportions.

Case Study: Pattern Recognition to Ensure Normal Operation

One of my favorite experiences with pattern recognition was at the Exxon Mobil Baytown chemical plant. On the balers used to package synthetic rubber, the control panels contained many different rotary switches that could each be set in either of two positions. In the original configuration, the orientation of those switches was standard: The left represented "off" and right represented "on." Other settings for switches unique to balers had no such industry standard. For the switch used to select either the main or the auxiliary hydraulic unit, the left represented the "main" unit and the right represented the "auxiliary" unit. As a result, the control panel had a random appearance; some switches pointed to the left and some to the right during normal operation.

One of our electricians decided to mistake proof the setup of a baler by rewiring all the switches so that in normal operation they all pointed to the right. He created a pattern where none had existed before. From then on, the successful setup of a baler was made more certain by pattern recognition. Even from the feed conveyors 20 feet away, it was apparent that when all the switches pointed up and to the right, the baler was set for normal operation.

Case Study: Creating a Pattern for Complex Piping Arrangements

At Exxon, we have developed a capability for pattern recognition on our complex pump stands where several pumps can be in service in different combinations or lineups or where a standby pump can be either in or out of service. It is normal industry practice for the handle of a quarter-turn ball valve to be aligned with the pipe when the valve is open and for the handle to be "across" the pipe when the valve is closed. Although this practice provides good visual information about the status of each individual valve, it does not provide any information about the status of the pump stand as a whole. To rectify this, we allow operators to change the convention *if it improves their work.*

The permissible alternate convention allows the operators to install the valve handles so that, in normal operation, all the valve handles align with the pipes, regardless of whether the valve is open or closed.

Note: When this convention is adopted locally, we are careful to mark the valves so that the status of each individual valve is always clear.

In this way, the operators can create a visual signal that the alignment of all valves is "normal." This is the hydraulic equivalent of making all the switches point to the right during normal operations.

The critical consideration in all of these situations is that we provide operators (or, even better, operators provide themselves) with a clear second chance to produce perfect work by eliminating errors through the simple expedient of creating patterns that make their work or the state of their process visually apparent. Operators who have done the work, along with anyone else who is interested in observing the work, can easily look at the visual indicators and know that the work has been done correctly. The real fun of pattern recognition occurs when you and your frontline teams are able to modify the work or the workplace so that everyone can easily recognize situations where things are not as they should be. Then, the teams can intervene with assurance and speed to prevent consequences.

Poka-Yoke *Practice 4: Simple Physical Devices and Other Minor Changes*

The fourth type of *poka-yoke* practice is creating simple physical devices or making minor changes to the work that allow the operators to be more successful. Once again, this is something with which we have a lot of experience in our personal lives. The key is to teach your operators to apply what they already know to their work.

Case Study: Adapting What We Know to the Workplace

One such device that we have adopted at Suncor is part of our common experience at banks or restaurants with canopy-covered drive-through service windows. Our base plant at Suncor is more than 40 years old and some of the pipe racks and electrical cable trays that cross over our in-plant roads are not as high as we would make them today. In order to protect these fragile systems from collision with the large vehicles common in our plant, we have adopted the practice of flexibly suspending a horizontal pipe across the road with dangling chains located at an elevation about 6 inches lower than the cable tray. Any vehicles that are too tall to clear the chains get a clanging reminder not to use those roads. Our prior practice of posting signs experienced several failures, but the chains work every time. They give a driver who has made the mistake of driving past the signs a second chance to avoid the bad consequence of driving into a pipe rack.

Case Study: Add a Small, Extra Step to the Work

When I first arrived at Suncor, I lived alone for some time in a rented condo. As a person accustomed to living in South Texas, I found the weather in northern Alberta quite cool (–55°C during the winters), so I frequently began wearing flannel shirts. After shrinking several of them in the clothes dryer because I forgot to change the temperature setting, I developed a small *poka-yoke* device of my own. I taped a piece of cardboard onto the dryer face in a way that it would hinge out of position when the dryer door opened and would fall back into an interference position once the door had opened. In this manner, it would prevent the door from closing until I manually moved it. This always reminded me to check the dryer's temperature setting.

By creating this small additional step, I interrupted the natural flow of the work and, in *poka-yoke* terms, I separated the mistake—forgetting to set the temperature properly—from the bad outcome—shrinking my shirts. This was not a physical separation, but rather a separation caused by the small new task that separated the mistake from the outcome. It was completely unimportant as a work element, but it caused an effective break in the action that was sufficient to prevent mistakes.

As you deploy the concepts of *poka-yoke,* you will certainly find many similar examples of zero-cost and zero-effort devices or work elements that can be implemented to prevent all manners of mistakes.

Case Study: Resequence Work Elements for Added Safety

At Exxon, we required that cars be *backed into* their parking spaces. We discovered that most accidents occurred as vehicles *backed out* of parking spaces into the line of traffic or into the path of pedestrians. Because there are rarely other moving vehicles or pedestrians in parking spaces, by changing the way people parked in this simple manner—putting the least capable activity (backing up) into the safest part of the sequence (backing in)—we greatly reduced the risk of an accident. None of the work of parking a truck and driving it away was changed, but by introducing a modest resequencing of the work, we almost completely eliminated on-site auto accidents.

APPROACHING PERFECT PRODUCTION

As you will recall from the discussion of statistical process control, mistakes are among the most common sources of special cause variation. *Poka-yoke* is a powerful companion to statistical process control that enables you to approach perfect production. I use the word "perfect" to describe the outcome of work where the process is capable and where the operators and others get a second or even a third chance to do the work without mistakes. Perfect work is indeed a legitimate result of this effort and a clear goal of lean practice, especially when lean practice is combined with statistical process control, continuous process improvement, and mistake proofing.

As we eliminate mistakes, it becomes increasingly possible to focus on improving the process capability of the production system, and, as the system becomes more capable, it is easier to focus on avoiding mistakes. In combination, you should begin to produce amazingly good performance that, at the very least, meaningfully approaches defect-free production.

These four common forms of mistake proofing will give your people a better chance to get their work right. People can physically separate the mistake from the consequence. They can make perfect (or imperfect) work visually apparent to themselves and others using colors, shapes, textures, sizes, and any other visual indicator they can imagine. Technicians can use existing patterns or create easily recognized new ones that allow them to be certain that the work or the equipment is as it should be or enable them to respond when it is not. Finally, you or your frontline teams can make modest changes to the work or introduce simple devices in a way that will create new opportunities to do it well. Many different types of potential *poka-yoke* practices or devices are simple enablers or reminders to do the right thing in the right way. You and your crew can try as many as you like with real certainty that each one will help produce improved results and none has the potential to cause harm.

Key idea: As you can see, mistake proofing is a technique you can promptly deploy and that your teams can practice with a high degree of autonomy. By applying the concepts of mistake proofing, people are not substantially changing the equipment, the product, or the process. They are simply ensuring that the work they do is *always* done without error.

9

Equipment Reliability and Operator Care

INTRODUCTION

In order to progress toward the lean value of operating in a way that allows you to produce any of your products at any time in any quantity and *always* with good quality, it is *mandatory* that your equipment be sufficiently flexible, capable, and available to support your efforts. We have addressed the technologies that will improve the flexibility and capability of your plant; it is now time to focus on equipment reliability and availability.

FINDING THE CAUSE: SEPARATING THE PROCESSES FROM THE EQUIPMENT

Reliability is another area where the process industries generally have both a greater problem and a greater opportunity to benefit from lean tools and thinking than do our mechanical manufacturing counterparts. Once again, this is a result of the differences in capital intensity between liquid industries and mechanical manufacturing.

Driven largely by the complexity of our processes, we are more reliant on technology and the equipment that enables it. Our equipment is generally larger and more integral to the manufacturing process than equipment in comparable mechanical shops is because the latter rely more heavily on manual labor. I have long believed that mechanical manufacturers generally produce complex products through relatively simple processes and that chemical manufacturers generally produce relatively simple products through complex processes. Our equipment is often the physical manifestation of those complex processes.

Key idea: Because the equipment in liquid manufacturing is often such an integral part of the process, many operators and even engineers believe the equipment is also complex in a manner that corresponds to the complexity of the process. This usually is not true. Although the reactive processes of chemical manufacturing often are very complex, their complexity is not typically an attribute of the machinery. Even in the most complex chemical reactions, the machinery is often just an aggregation of straightforward, engineered devices.

Case Study: Solve Operating Problems by Correcting Equipment Problems

The "Froth Density Team" at Suncor provided one of my favorite examples of this concept. As the bitumen is separated from the oil sands, the first intermediate product is a froth of bitumen and boiling water. When we recently accelerated our production rates, it was believed that a critical limiting factor was our inability to exercise direct control over the density of this bitumen froth because of naturally occurring variability in the oil-to-sand ratio of the ore that we mine. Therefore, we launched a froth density improvement effort. After 2 months of serious analysis and activity, we had completely resolved the operating problem by correcting deficiencies with our equipment. After careful analysis of our complex chemical process, all of our operating problems were routine mechanical problems. The resolution was not associated with the chemical complexity of the process in any manner.

The Role of Equipment Reliability in Lean Practice

Reliable equipment is a necessary characteristic of lean manufacturing, but most liquid industry plants (and most mechanical plants) are sufficiently unreliable initially that it is impossible to practice lean operations successfully without substantial improvement in the reliability of equipment. The fundamental reliability of liquid processing equipment can be improved through serious engineering; however, chemical plants generally have technical capability to achieve that. Certainly, reliability engineering is necessary to success, but it is not likely to be a factor that will distinguish your plant from your competitors'. Therefore, just as we have done with other technologies, this chapter will focus on the ways that equipment reliability can be enhanced through the engagement and autonomous action of your frontline teams.

The practice of engaging many people in equipment care is variously described as *owner-operator* maintenance, *operator care* maintenance, or *autonomous* maintenance. Each of these descriptions has a different connotation. We want the *operators* of our equipment *to act* as if it is their personal equipment, *as if they are the owners of the equipment.* We want the people who operate the equipment to provide the basic care and service that are within their capabilities in a way that allows the technical staff to move on to more complex activities that only they can perform. We also want the operators at the frontline of the business to operate, maintain, and improve the equipment autonomously in order to bring out their personal-best contributions.

OPERATOR CARE

As with many other lean practices, the foundation of successful lean maintenance is enabling broad participation. In this case, widespread participation implies that the people who operate the equipment will help to care for and improve it. *Owner-operator care* improves reliability as a direct result of new contributions from the operators. Furthermore, operator care is an enabling element that allows engineers and other technical specialists to improve their best contributions.

Key idea: When operators are providing basic care, engineers can pay closer attention to the big events and critical equipment programs that only they can achieve.

In this way, as with all lean practice, a synergistic relationship soon develops. Because the operators make their *new* contribution, the engineers' *existing* contribution can become better. When you create and use this new capability, it can make your operation notably better than and distinguish it from the operations of your competition.

There is nothing terribly sophisticated about either the concept or practice of operator care. With the exception of a few very wealthy people, we have all, at least to some extent, personally adopted the practice of taking care of the things that we own. Within the limits of our time and abilities, we provide basic care and maintenance and perhaps some improvement. We monitor the condition of our possessions and, when maintenance is required that exceeds our capabilities or time, we "represent" our possessions as we describe the problem to service professionals. When we take care of our possessions ourselves in this manner, they require less professional service and expense. We prolong the life of our assets and they provide us with greater enjoyment and value.

In our plants, each of us has also experienced individuals who bring that ownership attitude to work with them. Within the limits of time and resources that we impose on them, they attempt to provide the basic maintenance elements of care, monitoring, and representation for the equipment they operate. At work, as at home, when basic care is routinely taken, less intervention from skilled professionals is required; when intervention is required, it is less intense and more effective and the life and value of the equipment are enhanced.

Key idea: As with many other aspects of lean manufacturing, the goal of operator care maintenance is to provide a formal mechanism so that everyone can routinely do what traditionally has been done informally by only the best few individuals.

In a very real way, rigorous operator care in our plants is not much more than a formal version of the care that each of us provides to our possessions or the care that we would provide if we had adequate skills, resources, and time. For this reason, we already know what we want. The secret to success is establishing operator care in a *formal and sustainable* way that reliably produces that outcome from all the people all the time. Creating a sustainable program of operator care requires developing specific expectations that must be met as a routine part of each operator's daily work. It requires that managers and engineers participate in the process by making the equipment more capable of being maintained by operators. It requires that management provide people with the skills, time, and resources needed to meet those expectations.

Although this is described as *autonomous maintenance* (meaning that the operators do it with little or no supervision), *operator care cannot be informal or voluntary* because it is only one part of an integrated system of equipment maintenance. As with other forms of autonomous action, this work *is not closely supervised* but it *must be carefully managed and reliably delivered.* We must receive a predictable outcome from the maintenance activities at the frontline to make this system function in a way that enables the nearly perfect equipment performance that lean manufacturing demands.

Taken together, these several requirements imply that implementing operator care maintenance must proceed at a deliberate pace so that, at each step, everyone involved has the time and ability to learn and experience the capabilities required to sustain playing his or her part well. In order to succeed, autonomous maintenance of our equipment needs to be based on a very well-managed discipline of practice.

The Fundamentals of Operator Care

When I am asked to describe operator care informally, I normally say that the essence of operator care is that the individuals or teams who are with the equipment all day every day in the normal course of business are responsible to *keep the equipment clean, cool, and lubricated; change the oil, change the filters, and pay attention.* Obviously, those particular activities are not precisely correct for every operation, but they allow people to understand the essential elements of the operator care task easily. In this chapter, we will use that list of activities to describe the steps that you can follow in developing an operator care program at your plant.

Phase I: Basic Care

The first elements of the basic description of operator care (clean, cool, and lubricated) describe the initial levels of service and monitoring. They are examples of the sorts of fundamental practices that you can probably use to initiate operator care.

You should plan to introduce operator care to your people at a level of intensity that allows them to initiate this fundamental practice with relatively little training and only modest deployment of additional time and resources to facilitate the work. If you begin in a way that requires too much from them or from you, the effort is not likely to be sustainable. These basic activities are a great place to start because they allow people to make an immediate contribution with only a modest start-up cost. Before beginning a detailed description of the work, let us run through the operating implications of keeping the equipment clean, cool, and lubricated.

For this analysis, "keep the equipment clean" describes the most basic element of engaging people with the machinery in a way other than operating it. As frontline teams assume responsibility for the cleanliness of their equipment, they begin to think about the machinery and to learn about it; they often begin to care about it for the first time. When people begin caring for their equipment, their attitude toward it is no longer the industrial equivalent of the attitude toward a rental car. The machinery that they now both operate and maintain becomes "their" machinery. That change in mind-set often makes a truly fundamental difference in the maintenance outcome.

Starting with keeping the equipment clean is a common opportunity because virtually all industrial equipment is "dirty" in some way that affects its performance and virtually all equipment offers an opportunity to improve some aspect of performance simply by engaging the operators to care about it rather than simply use it. These two things are always good initial efforts as you seek to obtain immediate value from operator care.

Key idea: Make sure that you initiate this effort in a meaningful way that produces recognizable value in operating performance. Simply initiating *a random effort* to clean unspecified or unimportant things *is rarely sustainable and rarely produces any value.*

"Keep the equipment cool" is the way to recall that the second valuable way to engage people in the care of the equipment that they operate is to have them *monitor* the condition of the equipment and *respond* to that condition. By assigning the frontline operating teams a responsibility for routinely monitoring and managing some aspect of the equipment condition—whether it is keeping the machinery cool, keeping it warm, or keeping it dry—you begin the process of engaging the team in considering and managing the condition of the equipment.

Engaging with the condition of the machinery as it operates is different from operating it and is different from keeping it clean, which is a static condition more or less independent of its operation. In a process plant, you are likely to find that monitoring the machinery is also quite different from monitoring the state of the reaction. Operators who are excellent at monitoring the process may act as if a sticky valve, a leaking flange, a failed lube oil pump, or similar equipment conditions belong to maintenance and are not part of their operations work. When process equipment is unreliable, unengaged operators treat it as a nuisance rather than as a problem to be solved.

Key idea: Monitoring equipment is a specific task that is much less intense in engaging people with the equipment than the final step of operator care, which is "paying attention." When people monitor equipment, they are assessing a particular, prescribed condition. Normally, monitoring specific conditions is an early step that can begin promptly because it is within the existing capabilities of your team. When operators arrive at the final step of paying attention to the equipment, we will want them to be independently able and willing to notice intelligently any anomalous operation in a way that requires real familiarity with the equipment and its performance.

The third basic element is "keep the equipment lubricated." Here, the concept is that, as you initiate operator care maintenance, the operators can begin to provide basic service. Therefore, start by finding some aspect of the routine, repetitive service your equipment requires and assign that task to the operating team. Lubrication is symbolic of such an activity.

Phase I introduces the operating teams to engaging with the equipment by initiating the three elements of basic maintenance: caring for and cleaning the equipment, monitoring the equipment, and servicing the equipment. We will now look at each of those elements in enough detail to let you get started. Getting started is always the hardest part, so a great deal of detailed description is provided for phase I. Future phases are described in much less detail because the more advanced practices are obvious extensions of the initial steps.

Keep the Equipment Clean

The most basic element of caring for equipment is keeping it clean. Most people know or believe that clean equipment runs better than dirty equipment. There is good reason for that belief. Using the concept of "dirt" to cover the many different foreign materials that can be in or on equipment and lead to a deleterious result, let us consider some implications of dirty equipment as they relate to equipment performance:

- Dirt on the exterior can infiltrate the working parts and damage or degrade normal functions.
- Dirty equipment can contaminate the product.
- Dirty equipment may be sufficiently difficult or unattractive to operate or service, so the operators and craft people who interact with the equipment do so in only a cursory manner that does not fully meet the needs of the equipment or the operation.
- Dirty equipment can hide important defects that need attention but do not receive it because the defect is not apparent or accessible under the dirt.
- Last, but not least, hidden defects in the equipment may be the root cause of extremely dirty conditions.

Key idea: Generic housekeeping of your factory is a necessary element of any respectable operation. The state of cleanliness and orderliness displayed in any workplace correlates directly to the safety and overall attitude of the people. You may or may not need to address basic housekeeping, but that is different from engaging people in the care of the machinery that they operate. However, it is important that your leadership team must remember that "keeping the equipment clean" is the first step in engaging people with maintenance of the equipment—not another step in housekeeping.

A very important aspect of keeping the equipment clean is that *dirty equipment is often dirty for a reason.* That reason is usually associated with an unknown or unresolved deficiency in the environment or the equipment. The environment in which the equipment operates may allow abnormal amounts of dirt to get into or onto the equipment. The equipment may also be deficient, allowing accumulations of oil, grease, product, or other materials that exfiltrate from the equipment due to leaks, cracks, or poor service practices. As environmental materials or fugitive materials from the equipment or the process adhere to the outside of the machinery,

FIGURE 9.1
Polypropylene fines (white) adhering to a ribbon blender.

they normally attract other deleterious contaminants. Either alone or in combination, these materials disrupt the performance of the equipment.

Figure 9.1 shows contamination inside a ribbon blender used in polymer manufacturing. As plastic granules flow through the blender, some of the granules fracture to become crystalline "fines," which naturally adhere to the blender as the result of static electricity. This material is not inherently bad and, if it kept moving, it would not cause any problem. It becomes disruptive when it stops moving and accumulates on the surfaces of the blender. There, the adhered material might create a contamination problem when changing products or might cause operating problems by dislodging in clumps or simply by aggregating to such a degree that it interferes with the normal operation of the blender.

Independently of what the "dirt" in your plant is and how the equipment happens to become dirty, when the accumulation of unwanted material detracts from operational performance, the operators can play a beneficial role by keeping the equipment clean. That provides a good description of the difference between keeping the equipment clean as an element of operator care and generic housekeeping. In this situation, you will be seeking to implement a change in cleanliness that directly corresponds to an improvement in equipment performance.

As you commence operator care maintenance, you also need to change some aspects of your professional maintenance activity to support operator care. As mentioned previously, operator care is only one part of a new, integrated maintenance practice. Three critical and distinct elements of this new practice now will be shared among the various participants in maintenance: cleaning the equipment, correcting the reason the equipment is dirty, and, finally, establishing a routine to keep it clean.

Step 1: Initial Cleanup

The first element of operator care is to make the equipment clean. In most plants, even those with generally good housekeeping standards, the equipment at the outset of your effort is not likely to be as clean as excellent operations need it to be. Before your operators can *keep it clean,* the equipment usually needs to be *made clean.* Even if it is reasonable to expect that in the future the operating teams will be able to keep the equipment clean as a routine part of their normal operational roles, it is often not reasonable to ask them to undertake the initial cleanup unless you provide them with specific time to do the work. This is likely to be new work for your frontline teams, so management will probably need to provide some safety training and some special materials that will enable them to do the work properly.

The initial cleanup is often quite difficult. In most chemical plants, it means recovering from years of operation when the equipment was not routinely kept as clean as you now want it to be. Operators have not traditionally been asked to keep the equipment clean in this manner and the maintenance staff traditionally has not had the access or resources to provide that service. This is not a bad situation. It is, in fact, the most common situation.

Key idea: The reason for the new effort to achieve uncommon cleanliness is that you are no longer satisfied achieving only common results from your operations. It is very important that you *recognize and communicate that the new standard for cleanliness is indeed a new standard of performance.* When leaders and others understand that you are undertaking this effort as part of establishing a new standard of performance, they will approach this task in the appropriate way.

The difficulty of the initial cleanup is a good indication of why it is rarely good practice for management simply to announce that all workers are expected to clean up their work areas and to keep the equipment clean from this day forward. Unless you have a simple process and an inherently clean plant, this initial step will always be much more difficult and expensive than you may think.

You will need to provide substantial support to your people as they undertake the initial cleanup. Toward that end, you should be careful to initiate the new standard of equipment cleanliness in an area where it will have obvious value to performance that will demonstrably support the time and cost required. You then can proceed to extend operator care at a deliberate pace throughout your plant in increments that will ensure that each step is a valuable and sustainable success before you move on.

Key idea: Autonomous actions at the frontline of the business are not something that managers can simply request and receive. Autonomous actions are often interesting new things that people will enjoy doing as a change from the routine. However, they are always easily recognized by your operators as new work. From the very beginning of any initiative such as this, management must demonstrate that this new work *has the same stature* as traditional work. This means that as people make their contribution by keeping the equipment clean, management also makes its contribution by providing the time and resources necessary to the task. *For both frontline teams and management, this is not an informal or voluntary effort.* Autonomous maintenance can only succeed if everyone reliably plays his or her role well, and that includes the role of managing the activity for success.

Keeping the equipment clean is work that most operators have not done before and, in many cases, they have not previously seen anyone do it regularly. When you explain why it is important to adopt this new standard, you will need to make a credible case in order to attract the operators to help you. Further, when you commence funding the work with the time and other resources required to achieve the outcome that you want, you will also need to be comfortable that it is a good and sustainable business decision. Therefore, you need to start in a place where enhanced operator care will make a difference and you need to proceed in a way that will receive the support required to sustain the effort.

Key idea: The two most common mistakes made by management when initiating a cleanup effort are (1) to start in a place where the work will be easy but unimportant, and (2) to start in the dirtiest place possible, independent of business importance.

Although starting with a small or easy effort appears to ensure the desirable outcome of a quick and sustainable success, you must remember that *the success you are seeking is* not simply clean equipment but rather *equipment that demonstrably performs better* because it is clean. An easy win at improving the state of cleanliness has little value unless your teams understand the value and sustain the operating improvement. Even making a visually big improvement in a *really* dirty situation has no sustainable value unless it results in improved performance.

In your initial efforts, you need to be very careful to get this selection right and commence in a way that demonstrates your intent to improve performance, rather than just cleanliness. The initial efforts are always watched closely by everyone and therefore play a significant role in communicating your intent to the organization.

Autonomous equipment maintenance normally commences with cleaning the equipment because cleaning requires less training and fewer other resources than the subsequent steps of engaging people to care for the equipment. People generally know how to make things clean, so starting by making the equipment clean enables them to engage relatively quickly and successfully.

Note: Ensuring that a sufficient business outcome will result from practicing the new standard is a management responsibility, rather than an operator responsibility.

Case Study: Correlating Cleanliness to Operating Performance

In the manufacture of synthetic rubber, the last step in the process is to compress the crumbs of rubber that leave the reactors into bales similar in size and shape to hay bales. At this stage of the process, the rubber crumbs are sticky, elastic, and subject to static cling. The equipment used to create the rubber bales is a large hydraulic press with the traditional attributes of such equipment: Grease and hydraulic oil escape from the machine and accumulate on and around the equipment. In combination, the hydraulic oil, machine grease, and sticky rubber crumbs traditionally create a real mess. Figure 9.2 is a close-up of the accumulation of rubber and filth on the surfaces that compress the rubber.

This mess caused two major problems. First, this aggregate of grease, oil, and dirty rubber adhered to the machinery in a way that frequently resulted in product contamination as the product touched the equipment or as the equipment deposited contamination onto the product. Second, the oil, grease, and rubber accumulated in the service pits below the machines

FIGURE 9.2
Close-up of rubber baler working surfaces.

until these became such horrible places to work that no one wanted the job of servicing a baler. In this condition, it was obvious to everyone that our rubber plants were frequently rate limited and frequently produced defective product due to baler conditions that resulted directly from problems with cleanliness.

One day, as we were commencing operator care, I walked around Exxon's chemical plant in Fawley, England, with the plant manager; we approached the rubber finishing building at about noon. The plant manager led me into a baler pit, where several of his senior managers were waiting. We sat beneath an operating baler and enjoyed a very proper British lunch served by two caterers in formal attire.

That plant was never again rate limited by baler failures and that plant never again experienced a product contamination event caused by a dirty baler. Of greater importance to the long-term operation of the plant, the plant manager had demonstrated in a very compelling way to his entire team that a new standard of equipment care existed that correlated directly to operating performance. Figure 9.3 is a detailed current view of the balers in that plant; it is quite different from the view presented in Figure 9.2.

Step 2: Fix the Causes of the Dirty Equipment

Although management must initially provide frontline teams with the capabilities that will be needed to make the equipment clean, the real test of management's commitment commences once the equipment has initially been made clean. Industrial equipment is rarely dusty, as warehouses might be, and good plants are rarely littered. Industrial equipment, especially in chemical and other process plants, normally has industrial-

FIGURE 9.3
Detailed current view of rubber balers.

strength dirt, including grease, oil, and chemicals. Chemical manufacturing equipment is generally "dirty"—not with dirt but rather with materials that require substantial effort to remove. More importantly, industrial equipment is generally dirty because of an operating condition that will promptly make it dirty again unless the source of the contaminant is found and corrected.

Frequently during the initial cleaning, it is discovered that the source of the contaminating material is a leaky valve, an eroded gasket, damaged pump packing, or product that has otherwise escaped from the process and adhered to the equipment. The operating team knows this and knows with certainty that the new standard of cleanliness will not be sustainable unless management promptly intervenes to correct the underlying causes of the problem. Correcting valves, gaskets, pump packing, product leakage, and most similar causes of industrial dirt is generally beyond the capabilities of the operating team at this stage of the process.

These things normally require skilled maintenance and providing that resource to the teams is a management responsibility. *Operators fully understand that this is a test of management commitment.* Quite literally, if you do not follow through now, your initiative will quickly lose momentum and you will return to your original state of maintenance. In all aspects of operator engagement, everyone has a role to play and this is yours.

That said, this process of identifying the previously hidden sources of industrial dirt is just what you were expecting, even hoping, to find. According to lean values, these are the underlying problems that have been accommodated by allowing the equipment to become dirty. Your lean intent is to remove the original problem and then remove the accommodation that has allowed the problem to exist. As in the example of the rubber baler, the same repairs that enable people to keep the equipment clean will also enable the equipment to perform better or contaminate the product less.

Key idea: Generally, the sequence for applying lean principles to the relationship between problems and the resources that accommodate them is first to eliminate the problem and second to remove the accommodating resources. In some instances, such as initial cleaning, you may know that problems are hidden under the dirt, but it may be necessary to remove the dirt first in order to know the details of the problem that needs to be corrected. That alternate sequence is fine. Critical to your success is managing the relationship between problems and resources in a way that fits each situation.

Finding and repairing previously unrepaired leaks and other defects in the equipment will require that you spend maintenance money or consume other resources that you have not previously used on this particular problem. When you find situations such as this, as you surely will, you need to be certain that you are prepared to fix the equipment problems that are uncovered by the cleaning effort. Autonomous maintenance, beginning with keeping the equipment clean, will never succeed or sustain itself if you demonstrate to your team that cleaning the equipment is important to you only so long as the other people are doing their part but is not important when it becomes necessary for you to do your part.

Key idea: After the initial cleaning, it will be necessary for you to have adequate maintenance resources to correct the newly discovered problems. Do not believe that your people will voluntarily continue to clean equipment that quickly becomes dirty again because of an obvious problem that you refuse or fail to correct. You may compel compliance for a short period, but the compliance model of autonomous maintenance is never sustainable. This is another reason to proceed at a deliberate pace.

Case Study: What to Do When a Solution Is Not Immediately Available

In Exxon's adhesive chemicals business, during the initial cleaning we discovered a problem with sticky polymer that escaped from the process and adhered to the outside of the equipment in a way that attracted all manner of filth. The filthy material on the outside of the equipment frequently interacted with the process to contaminate the product and, as the accumulations grew, they frequently interfered with the normal functioning of the equipment. Unfortunately, at the time of the initial cleaning, we lacked a technical solution to this problem.

In one of our plants, the plant manager and his technical staff commenced a highly visible and ongoing series of experiments to find a way to fix the problem, and over the course of about 6 months they succeeded in that effort with a lot of valuable input from the frontline team. Because it was always obvious that management was indeed doing its part, the operating team stayed with us. A return of product contamination and equipment malfunction was avoided by sustaining a high level of cleanliness throughout this period. In an identical plant, the plant manager simply attempted to enforce the new standard of cleanliness by management directive; the performance and product contamination returned to the original standards within a relatively short period, but were recovered when we transferred the improvement technology from the first plant.

Whether the identified need for repair is routine, like tightening pump packing, or something more exotic, like creating a new containment technology for an adhesive polymer, management will want to make a real effort to respond properly and promptly when the cleanup effort uncovers a problem. Therefore, remember to start slowly, advance in increments, and work in places where the improvement is obviously worth the effort.

Key idea: In exactly the same way and for the same reason that the initial cleaning is difficult, the *initial repairs are likely to be more extensive and more difficult than you imagine.* Giving yourself time and budget to do it right will be more valuable than rushing deployment throughout your plant. Only by doing it right will the effort be sustainable.

Case Study: Side Benefits

Once the new standard for the cleanliness of rubber balers became common throughout the several plants in the Exxon circuit, we began to realize

FIGURE 9.4
Current state of baler plant.

benefits beyond improved reliability and reduced product contamination. Our plants became so clean that our material, which had previously been used primarily as a component of tires, was approved as a packaging material for foods and medicines. Another product in this family that had originated as a window sealant was approved as a base material for chewing gum. We even had our plants certified kosher so that our materials could be used as packaging for those products. Figure 9.4 shows the current state of a plant that was once so dirty that no one wanted to operate or maintain it.

Step 3: Routine Cleaning

Once you complete the initial cleaning and implement the initial repairs, future cleaning and repairs ought to become part of the production routine to prevent your gains from slipping away. An important part of creating a sustainable effort is to develop and deploy clear standards for cleanliness and specific work practices for routinely achieving the standards. This includes establishing meaningful *measures* to demonstrate the state of cleanliness. More importantly, it includes new measures that demonstrate the cleanliness-related *performance* improvement. The ability to measure the improved performance that results from the cleaning is a powerful indication that the effort is important and a powerful incentive for both managers and operators to ensure that the effort is sustained.

In the synthetic rubber experience, we visibly demonstrated the increase in production rates that resulted from improved reliability as well as the decreased amount of product rejected due to contamination from oil and grease. When it later became apparent that our new standards had created opportunities to expand our product line into food and medicine packaging, everyone involved enjoyed the real satisfaction of shared success.

Key idea: It helps to discuss the dirt on your equipment as an accumulation that accommodates the existence of equipment problems. This lean-thinking approach offers a different way of looking at dirt that is likely to drive a different approach to the situation. Clearly, when you can put a production value on the outcome, the task is not housekeeping; it is maintenance, but *not the old way of doing business; it is a frontline contribution to achieving lean performance.*

Step 4: Share the Work

In process operations that are in production around the clock, four or five teams often operate the same equipment. In those situations, it is good to have a generic standard for the state of the plant and equipment at the time each shift changes. This ensures that all teams do their part and that the several teams have an objective means to discuss the state of the plant among themselves. It is also good to have separate standards for specific equipment that each team will specifically maintain at a level above the general standard. As the autonomous maintenance program progresses, the detailed responsibilities of each team's members will include other, more advanced forms of operator care for the equipment assigned to their team.

Commencing operator care in a way that presumes such detailed work in the future is a good way to start. Further, establishing a disciplined approach to sharing the detailed work is always a good idea when you have multiple shift teams that share the equipment. Having both general and specific standards as well as formally shared responsibilities makes it clear that the teams are working together toward a common goal, provides a focus for more detailed work by each of the several teams, and reduces the delays and confusion that might otherwise arise if all teams had to agree on the details of each initiative before it could be implemented. In this way, each team has its own initiatives as part of the whole, and all teams are more likely to accept improvements initiated by other teams because they have improvements of their own that they want other teams to adopt in return.

Key idea: Whether it is for purposes of maintenance or improvement, when you have several different teams engaged in the same process or equipment, it is valuable to allocate detailed work to each team formally. The goal is that each team will adopt improvements created by the others. This greatly accelerates the pace of change.

With clear standards and objectives and visible measures of the new expectations, operator care maintenance to keep the equipment clean is not a complicated activity. Keeping the equipment clean once it is clean and once the causes of dirt are removed becomes another routine part of the job. Management's role is to

- assign the general and specific areas of team responsibility
- define the standards to which the equipment must be maintained
- provide the safety standards, the tools, and other capabilities required to do the work
- keep the equipment in an appropriate state of repair that will allow the frontline teams to keep it clean. *This is absolutely required and at least initially only management can do this!*

In this environment, frontline teams can provide a valuable new form of basic care to the equipment. As you begin this practice, you will experience the inherent nature of autonomous work: *Management establishes a very well-structured and carefully managed framework within which the frontline teams can take independent action.*

There is an ongoing need for managers and team leaders to respond properly as new equipment problems or new cleaning initiatives expose new reasons for dirty equipment. Initially, relatively close management involvement near the frontline is necessary. You will want to start slowly so that you can sustain this focus as the program expands.

Key idea: The result of cleaner and better maintained equipment will be just what you wanted it to be; therefore, you should be as happy to pay the cost of this maintenance effort as you would be to pay sales commissions. In both cases, *people are doing what you initiated and you are paying a small amount for a large benefit.*

At this point, frontline cleanup does not seem—and is not—very autonomous. Initially, the critical outcome is to engage the operating teams with the equipment in a way that produces a sustainable benefit. As the practice matures, there will be more autonomy.

An Alternate Step 1: The 5S Methodology

Some managers report that they are genuinely unable to identify a part of their facility that they believe would benefit operationally from an enhanced state of cleanliness as described before. The disruptive operating problems are more in the nature of "orderliness" than "cleanliness"; that is, the work area performs poorly because it is cluttered and confused, rather than dirty. If that is your situation, then you may substitute what is generally known as 5S practices as the initial step toward operator care in your plant.

The 5S technique is based on five practices that all begin with the "S" sound in Japanese. These practices, when translated into English, are often described as *sort, set up, shine, standardize,* and *sustain.* A brief description of each step follows:

1. *Sort:* Review the working area to identify the source of clutter as represented by materials, equipment, or other items that have accumulated in the area, but that are not currently needed for the task at hand. The improvement step is to sort the contents of the work area and reduce impeding clutter by removing to another location all the items that are not immediately needed for the current work. This task normally includes:
   - returning excess materials to a warehouse
   - storing things in a local standby area if they are not needed now but may be needed relatively soon
   - moving materials to a centralized storage for items that will not be needed for a longer term or
   - scrapping or selling truly unnecessary items for permanent removal

   Removing materials, tools, and equipment that are not immediately needed makes the state and activity of the items remaining in the workplace more visible and, therefore, more manageable. Process plants normally accumulate things like spare pipe spools, blinds and blanks, and scaffolding. Shops accumulate parts, tools, and things that are waiting to be rebuilt *someday.* When we undertook this 5S review in our central maintenance shop at Suncor, we found that our technicians had been stepping over and around things every day that had not moved in years. Equally, in our operating areas, we discovered that our technicians had been working around scaffolding that was erected for a job completed years ago. Simply "clearing the decks" created a lot of improvement and engagement and convinced people that we were serious about improvement.

2. *Set up:* Setting up is essentially the industrial equivalent of "a place for everything and everything in its place." Following a sorting step to remove unneeded materials, tools, and equipment, the work area is further improved by organizing the remaining items so that, once again, the state of the workplace is more visually apparent and there is greater ability to access the work and to move conveniently around the work area. This often includes shadow boards for tools, posting job instructions for routine work, as well as providing very specific locations for larger items so that it is always obvious where an item should be and equally obvious if an item necessary to the operation is not available or is out of place. In the original configuration of one of our shops, every craft technician had a personal toolbox and a half hour during each shift change was consumed by moving toolboxes around. During 5S implementation, we converted to providing standard tooling in each work bay appropriate to the work of that bay. By specializing the work and the tooling in each location, the planning and conduct of maintenance were greatly facilitated.
3. *Scrub or shine:* Scrubbing or shining the work area is another element of making the work and the work area visually apparent. Tools should be distinctively marked to indicate where they belong and which team owns which tools. This ensures that tools are always available when needed. Hazards, safety equipment, and other items should be distinctively marked to make them obvious. This work also has a housekeeping aspect, but it is a highly focused effort with a goal of making the area more organized and the state of the work visually apparent rather than making the area cleaner. Cleanliness is often an outcome of this stage of 5S practice, but not the core task. The key issue is to keep the work area scrubbed of all things that do not belong, are inappropriately placed, or represent some form of industrial dirt (that is, dirt related to the work). The objective is to make the work area more highly organized and more visually apparent in a way that improves pride, ownership, and effectiveness.

4. *Standardize:* Every plant has done the first three steps of 5S many, many times—at least once for each important guest to visit the facility. The practice of standardization is to move past episodic random cleaning and organizing to reach a state of organization that is standard and practiced routinely according to a known and fixed set of expectations. Often this set of standard expectations exists at multiple levels: one for each team and at least one more to standardize certain aspects of the workplace across all of the teams. Establishing a 5S workplace makes the work more apparent and easier to conduct and manage. Standardizing the 5S practice makes that benefit and the work that sustains it more apparent and easier to conduct and manage.
5. *Sustain:* As always, the final issue is to sustain the gains. This implies at least two elements. First, the new state of visual operations, cleanliness, organization, and standard expectations needs to be monitored routinely by managers or others who are not part of the operating team to ensure with a "cold eyes review" that the existing standards are being met on a continuous basis. The second element is to ensure that as the people, equipment, and even the products or other fundamental components of the workplace change or evolve, the standards and performance evolve as well and in a way that continuously supports the operation as it exists at all times in the future. Standard practices are not intended to lock in current performance, but rather to serve as a stable basis for future evolution to an even better state.

The 5S practices have a lot of value in organizing the workplace and making the state of the work visibly apparent. I will not cover this in more detail because it is well documented elsewhere and because the practice is not noticeably different between liquid and discrete manufacturing. The key issue for these purposes is that 5S can serve as the first element of implementing an operator care program for autonomous maintenance at the frontline.

Keep the Equipment Cool

The next step in phase I of autonomous maintenance is to "keep the equipment cool," which is a placeholder *for monitoring the operation of the equipment and responding when the equipment is not as it should be.* You can apply this concept in any way that you believe will enable your own frontline teams to contribute. The important part of this step is to engage frontline teams further in the care of the equipment by asking them to become more aware of the state of their equipment and to begin to respond when it requires attention.

One of the most valuable contributions of owner-operator maintenance is something that everyone naturally does well at home and at work: noticing when the performance, sound, or appearance of the machinery changes (more about this concept when we get to the standard of "paying attention"). This natural human practice of noticing when change occurs is consistent with many of today's most sophisticated maintenance technologies. At their core, the most technically complex reliability practices of condition monitoring are often just very sensitive forms of detecting or assessing changes in the equipment. The people continuously engaged with operating the equipment can make a great contribution to detecting these changes.

At Suncor, we have found a clear correlation between changes and safety. More than 80% of all our injuries and 100% of our serious injuries occur when something about the job changes and the operators continue working without recognizing or responding to the change. Great value to both operating performance and operator safety will be gained by establishing a clear practice of teaching people to notice and respond when the equipment changes.

The full richness of general equipment condition monitoring through autonomous maintenance activities will come later, but the process starts with monitoring and responding to a few simple, well-specified, and *valuable* characteristics of the equipment. Until the people at the frontline engage with the equipment, it is surprising the extent to which they are able to ignore changes that occur within the plant that would be obvious to them in other contexts. The technician who, at home, might notice the quiet sound of small bubbles in his swimming pool, indicating that the water level in the skimmers is a little low, might easily fail to recognize or respond to a much louder sound at work, such as that representing an abnormal change in velocity of a steam turbine.

Over time, as the extent of an operator's engagement with the equipment increases, you will find that people begin to notice and investigate nearly every anomalous condition autonomously (the ultimate and highly valuable state that we call "paying attention"). In the beginning, though, you will want to initiate the practice of monitoring and responding to the equipment in a way that your frontline teams can immediately achieve with their existing capabilities. In addition, the practice should be conducted in a quite formal, well-defined program so that you are certain the frontline teams can achieve it and equally certain that you can manage it. The autonomous part of autonomous maintenance comes after the engagement part. You are still not there yet.

Along with the appropriate team leaders and technical support, management should formally designate specific attributes of equipment performance that are valuable to monitor and should assign people specific monitoring responsibilities.

Note: You should also teach people a range of appropriate responses that—by skill and authority—they are able to take if their monitoring detects that the equipment has changed or needs service. I believe that people today notice far more changes than they respond to simply because they do not have a clear way to respond.

Your frontline people already monitor and respond to changes in process conditions. Monitoring the equipment is a modest extension.

Design a Visible Factory

As lean techniques have developed, the concept of a *visible factory* has become widely accepted. One aspect of a visible factory that is directly associated with engaging people in monitoring their equipment is *marking the equipment to indicate the normal operating state or configuration.* Operating technicians can mark the instrumentation of their equipment in a way that visually designates the normal operating range for each instrument. In the case of an analog pressure gauge, for example, this marking might be as simple as a green arc on the gauge face to designate the normal range along with one or more arcs in a different color to indicate abnormal conditions. Similarly, an electrician might rewire a control panel so that in normal operation all selector switches point up and to the right (see Chapter 8) or the operators might install an arrow on the site glass of a vessel to indicate the normal level of a liquid interface.

Any of these practices allows an operator to monitor the performance of the equipment or the state of the operation at a glance. That visibility alone has great value as operators run and monitor the unit. Once the gauges have been successfully marked or the control panel realigned, it is immediately apparent to a casual observer when one of the operating parameters, even on very complex processes, is out of its normal range or one of the controls is not normally configured. In addition, as the operators conduct this practice, it results in really valuable engagement and learning.

When they commence this effort to make the factory visually apparent most people promptly discover how little they know about their equipment. It is common even for highly experienced people to recognize that they do not possess sufficiently detailed information to mark the gauges and instruments correctly on the first try. Each succeeding attempt to make the visual representation of the process more accurate comes closer to formalizing the proper ranges, and each attempt results in increased knowledge of the process and engagement of the operators with the equipment.

In exactly the same way and for exactly the same purpose as when you deploy the initial cleanup of the equipment, you will want to proceed with the practice of equipment monitoring in a stepwise and deliberate manner. When you begin to ask people to monitor the equipment and respond when it changes, you will need to be certain that you properly describe *what* they ought to monitor and *how* they ought to respond. In almost every situation, you will have to provide your people with some skills, tools, or assistance that they currently lack.

This implies some additional attention from your team leaders or technical staff to select the conditions to be monitored, the attributes of monitoring, and the appropriate or allowable responses. Managers will need to provide those capabilities to the frontline teams in a meaningful way at each step of deploying or expanding the program of equipment monitoring. Providing your frontline teams with the capability to do this new work successfully is another form of demonstrating to them that this new work is valuable to you in the same way that you are asking them to accept that it is valuable to them. Precisely as you did when you initiated cleanup, you should begin this practice in a place where it will have demonstrable business value. You can deploy it further as you learn more about the ways in which you will get value from this work.

Once again, you will want to stabilize and sustain this practice by establishing clear standards of performance for the monitoring and response effort. You will also want visible and objective measures to demonstrate the success and value of the work. In most cases, if you establish a successful program of monitoring and response, you can anticipate that the reliability of the equipment will increase. You can normally detect the effect of that improvement as longer times between equipment failures or as reductions in the frequency or intensity of required service. In chemical processes, a very real value is often available in product quality or production capacity when the process becomes more stable or more consistent. Any one of these potential measures that is appropriate to your work will demonstrate the value of the monitoring and response effort.

Keep the Equipment Lubricated

The third element of phase I is to have the owner-operators of your equipment provide some basic service to the machines they operate. In most industrial processes, "keeping the equipment lubricated" is a straightforward activity well within the capabilities of the people who operate it. In addition, lubrication is a good placeholder for the initial step of providing direct service to the equipment because it is often a high-value maintenance activity that can be initiated at the frontline quickly, with only modest training and supporting resources.

Note: Lubrication per se is not the issue. You can initiate the first step toward providing basic service in any way that best suits your particular needs.

Your goal is for your operating team to begin to interact with the equipment as caregivers rather than just as operators, cleaners, and observers. Toward that objective, you will want to identify a meaningful activity that provides direct service to the equipment and is within their capabilities or can easily be made to be within their capabilities.

As always, you will want to initiate direct service with equipment that is obviously valuable and with service that is obviously worth the effort. Once again, you will need to proceed in a stepwise, deliberate manner so that, as you establish each new increment of operator care, you are willing and able to provide people with the resources and technical support needed to be certain that they are successful. Finally, you will again need to sustain this effort with disciplined practice and meaningful measures of success.

Often a meaningful measure of success for operators just beginning to provide direct service is to demonstrate that the service can be provided at a time convenient to the team and to the operation, whereas traditional practice interrupts the operation at the convenience of the maintenance department. You can also point out that, in some cases, lubrication or other routine service is not accomplished as intended because scheduling it is too difficult. The benefit of operator-delivered service should promptly appear as less lost time due to outages for service and greater reliability as a result of more rigorous performance of routine service.

Phase II: Advanced Techniques

The first part of the litany for autonomous maintenance is "keep the equipment clean, cool, and lubricated." These are the most basic elements of engaging people in equipment care, monitoring, and service. The second part is "change the oil, change the filters, and pay attention."

With the owner-operator elements of "clean, cool, and lubricated" active in meaningful units in your operation, you will have begun to demonstrate operator maintenance and to engage your people in the practice of caring for the equipment rather than just operating it. The increase in the knowledge and interest that operators then enjoy will have ongoing inherent value in many ways in addition to the immediate physical value of providing the equipment with a new source of routine care.

As you continue the deliberate process of spreading the basic elements of operator care to new operations throughout your plant, you can begin to deploy the more advanced elements represented by the second half of the litany in the locations where you initially started this work. If you initiated operator care in the original locations because reliability improvement has the greatest value there, then expanding the practice to higher order work there also ought to have the highest value. If, for some reason, that is no longer true, you can decide to deliver more complex maintenance elsewhere.

Define Your Goals

The three goals to keep in mind as you begin to implement the more advanced elements of owner-operator maintenance are to

- provide the equipment with superior service that produces superior performance
- greatly increase the operator's knowledge of his or her equipment
- provide basic service through the efforts of people who have not previously participated in maintaining equipment performance

Note: Our goal is *not* to fill the operator's day with maintenance work or to reduce the work of the maintenance department. The goal is to maximize performance throughout the plant by distributing the maintenance effort most effectively among the several people who interface with the equipment. Some important aspects of maintenance can be done most effectively only by the operators who are with the equipment all day every day.

Change the Oil and the Filters

Changing the oil and changing the filters are easy ways to remember that the next step in operator care is to move up to more complex tasks, including providing direct service. However, you will again want to define the exact next step to be taken in your plant according to your needs and the capabilities of your people. At this stage of operator care, chemical plant operators may begin to service pumps, valves, and other equipment where the service required is a defined program of adjustment or a readily learned pattern of taking a device apart, replacing a worn component, and reassembling the device.

This sort of work has an almost instantaneously beneficial impact on many operator-caused or operator-influenced equipment problems such as using cheater bars on valve wrenches, running pumps dry, and neglecting lubrication systems or pump seal-oil systems. Operators who have seen the inside of the valves and handled the components have a much greater appreciation for the system they manage than they did previously. It is also common at this stage for the operators to begin thinking about maintenance as an element of unit performance and to consider the individual items of equipment as part of a production system rather than as discrete components. This understanding of the "system of production" is often in marked contrast to the normal view of professional maintainers who do not have the perspective of either performance or system when they are working on your equipment.

Case Study: Maintenance as Part of the System of Production

In one interesting situation at Suncor, as soon as the operators began to take responsibility for replacing pump seals and servicing the seal flush system, they noticed that the seals were failing prematurely for a reason that was easily corrected. It turned out that the maintenance staff was replacing worn pump seals using seals identical to the originals that did not provide a liquid path for the seal flush system to be effective. After years of disruptive failures on an important pump, the first operator who looked closely at the pump seal as part of the whole production system concluded, "This can't be right."

Although it is likely that any engineer or mechanic who looked closely at the whole system with an eye toward problem solving would have noticed the same problem, the fact is that the problem had not (and probably would not have) attracted the attention of an engineer. Apparently, the several mechanics assigned to the task from time to time had been too focused on replacing the seals to worry about why they were failing.

Another example of the benefit that comes from operators who participate in equipment service occurred in Suncor's upgrading unit. The first step in upgrading bitumen to create synthetic crude oil is to crack the bitumen into hydrocarbon molecules of much lower molecular weight. Until that happens, raw bitumen is passing through our pipelines as a highly viscous and sticky liquid. To accommodate this property of the material, we have a policy that valves in many of our pipelines need to be steamed for 15 minutes before they are actuated. When that is not done, it often results in failure of the valve actuator because the actuator is not sufficiently robust to work against an accumulation of bitumen in the mechanism of the valve.

Before we implemented operator care, when a valve actuator failed, the operating team simply turned the equipment over to maintenance. As we initiated operator care, we began to ask the operating team to isolate and remove the failed valve and to steam clean it offline before turning it over to maintenance, who were then responsible only for mechanical repair. As the operators began to see for themselves the sticky internals of the valve, and the effect of steam on the movement of the valve, that source of valve failure effectively disappeared from our plant.

Deployment of more complex service will follow the pattern that you have already established with the earlier initiatives:

- Pick an area and an activity that have easily demonstrable value.
- Support the teams with the skills, tools, and other resources they need to ensure that they are successful.
- Identify the intended outcome of the work and measure your performance and results.
- Proceed at a deliberate pace that is sustainable because the value derived from autonomous maintenance exceeds the cost and effort of the work and ensures adequate management attention as the process evolves.

Pay Attention

As a companion to providing more complex service, the activities of phase II are intended to engage operators to pay attention to the equipment under their control. This often requires that the owner-operator represent the equipment when professional service is required. That means the equipment operator should be able to describe the state of the equipment, its normal performance, and the aspect of its performance or operation that has changed. Good operators often color this description of the current state with examples of prior events and the service that was effective at those times. This is the sort of behavior that we all adopt individually when we take our cars for service.

Paying close attention to the many changes that can occur in production equipment and responding properly when they occur are the culmination of the earlier elements of owner-operator maintenance. It is only truly effective when people have been through the previous steps and have begun to think of the equipment as the equivalent of their own property and have already gained good knowledge of how it works, sounds, and feels. They have a close engagement with the equipment and are able to behave or respond with a meaningful awareness of it.

People are not naturally equipped to replace ultrasonic testing for the detection of hidden cracks in machine castings or for intuitively conducting an equivalent to statistical analysis of equipment performance trends over an extended period. They do have great natural ability to detect many sources of change in their environment, including changes in the equipment that they operate. The sight, sound, smell, taste, and feel of the plant and the equipment become extremely familiar to those who operate it every day. All those human senses are extremely sensitive and reliable indicators of change and the human mind is an amazing tool for comparing the current state of the plant to prior experience. Your operators can fill this role by learning to pay attention effectively to the equipment that they operate.

At this point, *autonomous maintenance genuinely becomes autonomous.* The operators retain the discipline of assigned routine tasks to keep the equipment clean and provide basic monitoring and service. They still have the discipline of the more advanced tasks of delivering higher-order services. Reliable delivery of routine maintenance work is a necessary part of the overall system for equipment care and it must reliably continue.

However, *paying attention* to the equipment is substantially different. There are no new specified attributes to monitor. Instead, the operators are charged to pay attention to all attributes of the equipment and to detect any anomalous conditions as they occur. They may know how to respond to their observations, they may learn how to respond, or they may simply recognize that they need to get someone else to respond. The important issue is that the equipment does not change in any noticeable way without attracting the attention of the operator and without the operator initiating some form of timely response because, as we all know with some certainty, prompt response often greatly reduces the severity of equipment problems. *Operators who pay close attention to their equipment provide a capability for prompt response that cannot be duplicated in any other manner.*

Case Study: Question Conventional Wisdom

As is usually the case when we engage new people to pay attention to the equipment, they begin by questioning conventional wisdom that has previously gone unchallenged. Since the beginning of oil sand operations in the mid-1960s, nozzles in the extraction units regularly become clogged with "fibers." Because the formation of oil, including bitumen, is associated with dinosaurs, the traditional wisdom of the business was that these fibers were dinosaur hair, and therefore they were assumed to be a natural companion to the bitumen. As we began operator care practices, one of the process technicians, when unclogging a unit, saved some dinosaur hairs and passed them to his wife who works in our laboratory. After the simplest analysis, it became apparent that our "dinosaur hair" was man-made—nothing more exotic than trash that blows into the mine and finds its way into our process. Since he communicated this news and we launched an initiative to keep trash out of and away from the mine, our experience with clogged extraction units has greatly diminished.

In combination, the several elements of operator care culminate in the practice of human monitoring for unexpected change. That practice is amazingly successful and powerful. When it works well, the impact on equipment performance far exceeds your expectations. In combination with engineering programs to create capable equipment and statistical programs to remove special cause variation in the process, the attention of operators and craft technicians to the details of equipment performance can ensure that the equipment consistently operates at its best. That combination lets you begin to aspire toward defect-free production.

Autonomous Maintenance as an Element in Improvement

It is often said that the people who run the process are the people who know the most about it. Normally, that implies that the people who operate the process are also the people who have the greatest ability to manage and improve the process. In mechanical manufacturing, where people physically handle parts and perform other forms of manual labor in activities of relatively low complexity, the flow of material is visually apparent and the sources of process disruption that offer opportunities for improvement are often apparent as well. In that visible and physical environment, manufacturing processes can indeed be readily improved by people who simply possess good observational skills and some critical thinking.

In chemical and process manufacturing, those enabling visual signals and physical contact are often absent. In our highly complex plants, the product flow is often completely contained inside the equipment and people rarely, if ever, see the material as it is in progress. As a result, the people who operate our plants frequently do not have sufficient knowledge or understanding of the equipment to allow them to manage or improve it successfully. Without new knowledge and skills that can be obtained by participating in operator care, most operators experience process problems that result from poor maintenance but cannot fix them. They can name the experience and recognize that it has happened before, but often that is the extent of their contribution.

As your plant progresses through the steps of owner-operator maintenance, your frontline personnel will gain the knowledge needed truly to engage with the machinery. They will understand that the complex chemical processes are performed within machines that are *just* machines. They will learn where and why the equipment fails and they will know what happens on the inside when they turn the valve handle. They will know precisely what process condition each marked gauge represents. The hidden process will be more visible and they will have gained much greater understanding of the needs and benefits of routine service. Doing all that will make them comfortably familiar with their equipment, just as we are all comfortably familiar with the equipment that surrounds us in our personal lives.

Through the practice of owner-operator maintenance, you will create a relationship between your people and your equipment. People naturally want to care about their operation. They want to know how it functions. They want to participate in the whole range of activities associated with operating, maintaining, and improving their plant. Although nearly every manager and operator wants those things, until you give them the capability to learn and experience the plant in a new way, they simply cannot do it.

With this enhanced knowledge of the equipment and operation, frontline staff will contribute more than basic care and service. They will do three things. First, of course, they will become better operators. As a team, they will understand the equipment much better and, as a team, they can begin to run the process in accordance with this knowledge. The increased visibility and hands-on familiarity that operator care maintenance services create will give your operators the same ability to contribute to a process manufacturing plant as the mechanical manufacturing workers enjoy because of the natural visibility and physical contact inherent in their more tangible operations.

Second, the operators can directly contribute to improved operation by personally improving the care that the equipment receives. They will perform required maintenance at a level of detail and frequency that cannot be duplicated by maintenance technicians. They will also pay attention to and represent the equipment when professional or skilled service is required.

When we take our cars to the mechanic's shop, few of us provide instructions as simple as, "Repair as needed." Instead, we attempt to communicate the full range of our knowledge. We describe what was happening when the breakdown occurred. If there were any gauges in abnormal positions or any indicator lights were lit, we mention that. If the same thing has happened previously, we describe what happened that time and what was done to return it to satisfactory service. Certainly, if we have noticed any contributing causes that might not be immediately apparent to the mechanic, we mention those. We want that same awareness and representation to be part of the conversation when our operators interface with our mechanical teams.

The third characteristic of people who are truly engaged with their equipment is their far greater capability to cause improvement. The people who operate our processes do indeed know a lot about them, but they know about the processes because we taught them to operate the plant and they improved on their original performance as they gained experience. When we also want them to improve the equipment in the plant or to improve the entire system of production, including both the process and the equipment, we must also give them new knowledge of the equipment. As they gain experience with equipment, they can build that knowledge into an improvement capability as they did earlier with their knowledge of the process. We must teach people to improve the maintenance and equipment of the plant in the same way that we originally taught them to improve the operation of the plant.

Case Study: Engagement and Improvement

At Suncor, we operate 400-ton heavy haul trucks to transport the oil sand from the strike face in the mine to the first processing units that separate the bitumen from the oil sand. As we began to engage our truck operators in the care of their units, they began to do three things:

1. They began to drive more slowly on long trips under load to prevent thermal tire wear. They recover by driving faster when not under load. As a result, Suncor now owns several world records for tire life in heavy haul trucks.
2. They began to understand the relationship between the quality of the mine roads and the life of the truck frames and suspensions. As a result, whenever the truckers see a pothole in a mine road, they report it promptly so that it can be repaired before the equipment is damaged. Because of these simple acts of awareness and care, our truck suspensions are lasting far longer than industry standards.
3. As the operators recognized that tire wear was a controllable function of heat and speed and that damage to the suspensions and structure of the trucks was a controllable function of the quality of our mine roads, they recognized a new opportunity. At the time, our 400-ton trucks did not regularly transport 400 tons in each load. The loads had been reduced to prevent tire wear and mechanical damage. As the mechanical performance of our fleet improved, we took advantage of the more robust operations to increase the loads on our trucks. In addition to routinely having more trucks in the mine because of improved reliability, each truck now moves more material in each trip.

I like those examples from the mine team because they resulted in performance that far exceeds that of our competition. However, I recognize that they are not process industry examples, so the following are two more that are clearly liquid industry examples.

Example 1: In processing oil sands, the first step after lifting the bituminous sand from the mine is separating the oil from the sand. In the nature of that process, we pump a lot of sand-and-water slurry through very long pipelines. The velocity of the flow is critical. If we pump too slowly, the sand falls out of suspension and plugs the pipeline. If we pump too quickly, the sand erodes the pipeline and we are forced to shut down to replace the worn spools of pipe. As we began to engage operators in making the process visible, one team began to "mark the gauges" of both the pipeline and the separation cells.

At that point, the team noticed that the process flow in the separation cell was measured in U.S. gallons per minute (USGPM) but the pipeline flow was measured in barrels per hour (BPH).

Those measures are different (1 BPH = 0.7 USGPM), but close enough that new operators frequently attempted to "balance" the flow in both systems by making the flow rates numerically the same. That practice resulted in periodic surges of flow either above or below the critical rates in order to return the system to liquid equilibrium. Many people could have resolved that problem if they had known about it; however, an operator closely engaged in making the process visible was the first to understand the integrated system and to take formal steps to ensure that this misapplication of equipment would never again disrupt our operations.

Example 2: During one of our outages for major maintenance, a contractor mistakenly installed a carbon steel pipe spool in a stainless steel line used to transport the carbonic acid that results from scrubbing carbon dioxide out of our manufactured hydrogen. Because of the extremely cold winters in northern Canada, the next step after installing that replacement spool piece would have been prompt installation of heat tracing and insulation. In effect, the mistake of installing the wrong spool would have promptly disappeared under insulation until some later time when a catastrophic release of high-pressure hydrogen surely would have occurred.

Fortunately, an operator who was paying attention to the equipment noticed that there was one black pipe spool in an otherwise silver line. Although he did not fully understand the nature of the problem or the likely consequences, because he was engaged with the equipment, he noticed that the pipeline was visually different from what he expected it to be. He promptly got help that successfully recognized the nature of the problem and corrected it before there were any consequences.

AUTONOMOUS ACTIONS

The management role in creating and leading autonomous improvement is described in Chapter 10. However, if you are deploying the capabilities for improvement at the same time that you deploy the owner-operator maintenance practices, you can anticipate that your teams will use their new knowledge of the equipment as part of their tool kit while they practice improvement.

Autonomous maintenance and autonomous improvement share the same attributes. People have clear performance goals. They have the time, skills, and resources needed to achieve those goals. They also have a framework for action that defines the limits of practice in a way that ensures that their work will be well accepted and safe. The framework for action described in Chapter 10 also makes the practice of autonomous maintenance and autonomous improvement visually apparent so that managers and engineers do not need to intervene in the activity in order to be certain that people are doing the right things in the right ways within the right limits. Operator care maintenance is an essential component of the best practice of autonomous improvement in the process industries.

Case Study: Autonomous Maintenance Improvement

Figure 9.1 illustrates polypropylene (PP) fines adhering to the ribbon blender causing operating problems and product transition problems. One of our engineers brought in a cleaning system that was similar in concept to a sandblasting unit, except that it used dry ice as the blasting medium rather than sand. The benefit of dry ice is that it sublimes as the temperature rises and leaves no residual material to contaminate our process. Dry ice blasting turned out to be effective for this cleaning task, but it was expensive and relatively difficult to set up and use. Therefore, although it was effective, it was not adopted.

However, the use of dry ice as an alternative blasting medium gave one of our operators a good idea. He filled a traditional sand blaster with PP granules from the current production run and blasted the ribbon blender using PP granules as the abrasive material. This too worked just fine and it provided a successful, fast, and inexpensive cleanup of the blender (see Figure 9.5). The PP granules did not disappear as the dry ice had done, but they did not need to. Because he was using granules from the current production run, he was merely reintroducing material that was supposed to be there.

This operator was providing maintenance service to the unit based upon a good understanding of the full system of production. As a result, he was able to create a new solution that was uniquely effective. At that point in the evolution of operator practice, he and his team had gained the authority to practice autonomous maintenance and autonomous improvement; therefore, he did this on his own initiative and sustained the practice by teaching it to his colleagues.

Once autonomous improvement commences, it can grow quite quickly, often by simply reproducing an improvement in every place where a similar

FIGURE 9.5
Cleaning with PP granules.

opportunity exists. At Suncor, we had the same physical problem arise in two distinctly different operating situations. In both instances, it is unlikely that an engineer or manager would have looked at the problem faced by the operators in a way that would have recognized the true nature of the problem. Even if an engineer or manager had been involved in improving either of these situations, we would likely have enjoyed only one improvement because it is probable that the same engineer would not have been engaged in both situations and certainly not in all instances of both.

On the other hand, the operators, who were empowered to make autonomous improvements in certain situations, not only recognized the first problem and fixed it, but also recognized the same situation when it existed in a different way in another part of the same plant. Of even greater benefit, they then made the same improvement whenever the same situation occurred throughout the plant.

Case Study: Piping Designs—A Different Point of View

In order to service our large heat exchangers, it is necessary to empty them before they can be opened. Our operators discovered that the engineer who had designed the drain lines had used 2-inch pipe for the lines, but had supplied only a half-inch pipe fitting to connect the drain line to the waste truck. Therefore, draining the large volumes of water needed to service the exchanger often required several hours. However, there was no reason to have a small fitting on the end of a large pipe. It probably just represented the engineer's opinion of the appropriate fitting to connect to the waste truck. Perhaps at some time in the past that had been correct.

Fortunately, the operating team that made this discovery consulted with its engineering support and, with technical approval, its members took it upon themselves to replace the small fittings with 2-inch fittings. In that mode, the exchangers could be properly drained very quickly, greatly reducing the service outage needed to perform maintenance.

Note that the team did not have authority to alter the piping design autonomously without engineering consultation, but it obtained that permission in a way consistent with our management of change procedures. With engineering approval, team members then autonomously carried out the execution of the work. More important, they not only altered the exchanger that had originally attracted their attention, but also altered all exchangers that demonstrated this same condition. The detailed knowledge and exposure that are uniquely available to operators provide the ability to extend a solution to all similar opportunities on a complex unit. This is another way that the frontline teams can add great value to the improvement effort.

When we return pressure vessels to service after major repairs, we hydrotest the equipment to ensure the integrity of our repairs under operating pressure. This testing is done by pumping the unit full of glycol and pressurizing the entire system using a pressure and vacuum (P&V) truck. Operators of P&V trucks who perform this service normally have a centrifugal pump for high-volume pumping and a positive displacement pump for high-pressure pumping.

Here we experienced the same problem as that with the heat exchangers, but in reverse; that is, the problem was getting liquid into the system, rather than getting it out. The engineer who designed this system had again provided a 2-inch pipe with only a half-inch fitting to fill and pressurize the units. With that limitation, the P&V operators were required to fill the units using the positive displacement pump in a way that required more than 20 hours to fill a large vessel, although less than 1 hour would have been needed with a 2-inch fitting and a centrifugal pump.

The operating team involved in modifying the heat exchangers recognized the problem immediately as the same as the one it had faced before and promptly intervened (again with engineering support) to install new 2-inch fittings. As team members had done with the heat exchangers, they corrected all the pressure-testing connections in the units. In this manner, many problems were solved before they were ever experienced and we received many separate improvements arising from a single idea.

Our operating teams have plenty of capability and resources to change pipe fittings. In these situations, a quick ad hoc review with an engineer confirmed that there were no issues with authority. They were not changing the work or the outcome in any way. They simply made a few modest changes that allowed the work to proceed much more quickly. Chapter 10 contains a complete discussion of how operating teams acquire authority for autonomous action.

What impresses me about these stories is that the operators were not victims of the poorly designed equipment, as they previously would have been. Because they were adequately engaged with the equipment, they recognized the nature of the problem and that they could fix it. Once the team on the unit had the idea the first time, it was rapidly able to use it many times, each time with complete success. Following an ad hoc engineering consultation, the team did not have to call someone else to help. Its members knew what to do and just did it.

It has always been a pleasant surprise to me just how much improvement is possible when people truly engage with their equipment. You can and should expect to be happily surprised as you adopt owner-operator maintenance in your plant. It is truly one of the great lean tools.

10

Lean Leadership and Ethics: Creating an Engaged Workforce

INTRODUCTION

Leadership and ethics comprise the first of two Shingo elements that are described as the ***cultural enablers*** of lean. The second cultural enabler is *people development,* which will be described in the next chapter. Conceptually, leadership and ethics are both enormous topics. However, for these purposes, it is practical to reduce the broad topic to ensuring that leadership is sufficient to the successful practice of lean manufacturing. More specifically, the discussion comes down to ensuring that leadership creates the new industrial culture discussed in Chapter 1. In order to practice lean manufacturing successfully, you will need to design and create within your business an industrial culture where each person can make his or her personal-best contribution and the people at the frontline are enabled to make autonomous small event improvements.

The ethics of engagement are a close companion to leadership for this purpose. The industrial culture to which you aspire includes significant new elements of social interaction, teamwork, human diversity, and intentional alignment for both small teams at the frontline of your business as well as for the "big team," which consists of the entire enterprise. In most businesses, as you undertake this effort, it will be the first time that those social issues have been a formal part of your management consideration. As with all social issues, these are sensitive, personal, and important to the people involved. Managing this aspect of transforming your business will be both new and challenging.

Further, the new way of working must be fair in fact and it must be perceived to be fair. This is not a time to exploit the goodwill of your employees, but rather a time to work together to achieve the new state of success for the business you share with them. Many people will prosper in the new culture, but a few will not be able to tolerate the transition. They, too, must receive obviously fair treatment. The leadership and ethics of transforming the culture of your business are a serious undertaking.

That said, the most exciting aspect of lean practice is that the lean values and enabling technologies are all purposely directed to engage people at the frontline and others throughout the enterprise to improve your business by making contributions that only they can make. The leadership and ethics required to achieve that outcome offer a great reward. Our factories are filled with opportunities for improved performance that can never be brought to fruition in any other manner, as well as problems that can never be solved in any other manner. Capturing those opportunities in ever increasing numbers and enjoying the resulting improved performance are the most exciting things that you and your employees will ever experience.

The new excitement at the frontline extends to the managerial and professional staff, who previously bore almost sole responsibility for improvement. As frontline teams create stable operations and a mass of meaningful improvements, the engineers in the organization become more capable of creating improvements that only they can achieve. For engineers whose professional lives have previously been limited to industrial fire fighting, I have heard the new ability to focus on true engineering work described in terms ranging from "a breath of fresh air" to "hope for the future." This truly will be an exciting time as your staff begins to work at their full potential in the jobs they always intended to have.

Key idea: Henry Ford once observed that he could divide his career into three stages, which he related to his understanding of engineering. As he began his career, he believed that an engineer was a person who could do in 1 day the same things that anyone could do in 10 days. As he formed his business, he believed that an engineer was a person who could do for $1 the same things that anyone could do for $10. When he became truly successful, he understood that engineers could do things that other people could not do.

Ford's understanding is accurate: Your engineers can and should do things that only they can do. However, before engineers can make those special contributions, the operation must be inherently stable and the special causes of variability that disrupt production must be managed at the frontline. The simple fact of life in process manufacturing is that operational stability *must* be created and sustained in the process, and special causes of variation *must* be resolved. If they are not resolved at the frontline, the engineers, of necessity, will be called to that service at the expense of other work that they could be doing. *Initiating engagement at the frontline enhances performance in all parts of the business; it especially enhances the contribution of your existing improvement team.*

IMPROVEMENT EXPERIENCES AT THE FRONTLINE

Initiatives to gain the engagement of the frontline teams are often based on the belief that the people who run our plants know more about the daily details of the operation than engineers and managers do and, further, that they have the ability to act on that knowledge. The technicians at the frontline *do* have knowledge, desire, and capability that can allow them to operate the plant successfully and improve performance; however, today they cannot effectively act on that knowledge in most cases. The best proof of that assertion is that the intelligent, hardworking people in our plants are not currently making the sort of improvement that we need. They are good people, who are motivated and enthusiastic about making the businesses we share more successful. If they could contribute in a new way to make the business better, at least some of them would have already done it.

Statistics from the American Productivity and Quality Center in Houston report that the average rate of engagement for North American industry is one suggestion per year for each seven employees—only 20% of which are implemented. That results in a North American industrial average of only 0.028 implemented improvements per person per year. The North American result is about the same as the outcome for Western Europe. When you compare that average result with the performance of a truly world-class organization such as Exxon Baytown, where the corresponding number exceeded 40 implemented improvements per person per year in 1997, it is easy to recognize that a structural difference must be responsible for the result.

THE STRUCTURE OF EMPLOYEE ENGAGEMENT

Management is responsible for the structural results of any organization; creating an engaged workforce is indeed solely a management responsibility—a very rigorous responsibility. Although many managers still believe that if we just get out of the way, people will respond with the sort of improvement we need, that simply is not true. In fact, permission to take random or undefined actions is far more likely to result in chaos than in improvement. I would never work in a chemical plant that did not maintain strict *management of change* controls.

Key idea: To create an engaged workforce, management *must* create an organization in which most people can autonomously improve the business and, at the same time, in which everyone can be certain that all changes are the right things done in the right way and well within precise technical boundaries for management of change.

THE ELEMENTS OF ENGAGEMENT

The organizational capability for rapid but disciplined improvement is created when management delivers to people throughout the organization the *five objective elements of engagement:*

1. *Goals* to be achieved
2. *Skills* to achieve those goals
3. *Time* to practice improvement
4. *Resources* to cause change
5. *Framework for action* to ensure careful *management of autonomous* work

These are called the objective elements of engagement because it is possible to determine with some certainty whether they exist. In the vast majority of circumstances, when teams at the frontline have these objective attributes in place, they can and will initiate a successful practice of autonomous improvement.

Two subjective characteristics of engagement are less apparent. You will become most interested in the subjective elements when teams possess the five objective elements of engagement and yet have not engaged. In that circumstance, one or both of the subjective elements are stopping the team from succeeding. These subjective elements of engagement are

1. *Lack of trust* in management
2. *Interpersonal disruptions* within the team

In this chapter, we will examine each of these. Taken together, management of all seven attributes is necessary for people at the frontline of your business to engage in autonomous improvement. Managers cannot simply get out of the way. We need to create a new social culture proactively at work in which a new form of work can be effectively practiced.

Key idea: Creating a meaningful capability for autonomous improvement is an absolute necessity for world-class performance. It simply is not possible for a plant to achieve 40 or more implemented improvements per person each year without providing people with the ability to make improvements on their own initiatives. It is equally impossible for a plant with a traditional level of engagement to compete with a plant that is benefiting from a truly engaged workforce.

Clear Goals

We have previously discussed the three-level-view process for translating goals throughout the organization (see Chapter 3). There is no need to revisit that discussion except to emphasize that engagement at the frontline of your business cannot possibly succeed if people are uncertain about what to do. *There is no room in an engaged workforce for random improvements or uncoordinated change.* Every action taken by every team must add to and be compatible with the work and the changes of others. The principal mechanism for achieving that mandatory alignment throughout a large organization is goal or policy deployment.

People need to understand the unique contributions to the business that are within their independent capability to deliver. Of equal importance, people need to have a clear understanding of the actions they are not authorized to take independently. Good goals are both engaging and limiting. People also need to have sufficient awareness of the things happening in other teams and in other parts of the enterprise to ensure that they sustain the original strategic alignment and tactical coordination over time. The actions of all teams will evolve as the business evolves and as the state of engagement matures, but the strategic direction that aligns all the teams needs to remain constant and visible.

Skills Necessary to Achieve the Goals

Efforts to engage people at the frontline generally begin with great trust in the knowledge and capabilities of the frontline teams. Recognition of that inherent capability is well justified and amazingly valuable. The difficulty in progressing from knowledge of operations at the frontline to implementation of successful initiatives to improve operations is that knowledge of the opportunities is not the same as the ability to cause improvement. This limit applies even when the problem is an operational detail that appears to be well within the control of the people at the frontline.

There are two classic examples of this dilemma. The first occurs when the manifestation of the problem known to the frontline team is not the root cause of the problem that the team is experiencing. In this case, people know that a problem exists, but they do not know why it exists; no amount of activity directed toward the superficial attributes of the problem will cause the required improvement.

Case Study: Teach Your Teams to Look for the Underlying Cause

Suncor brings most of our people to the site each day in a fleet of chartered buses. The first thing that people do as they disembark is to swipe their identification badges on a card reader that automatically records the time they arrived. After this automated timekeeping system was initiated, many people became frustrated and even angry because up to 60% of all payroll entries were still done manually. The payroll governance team spent months training and retraining people how to use the system and badgering frontline supervisors about the importance of using the system. None of that made any difference. Everyone at the frontline understood how to use the system and supervisors greatly preferred eliminating manual data entry.

> However, because the buses belong to management, when the bus arrives late, we pay people as if they had arrived on time. That is, when the buses arrive late, each person on the bus requires a manual timekeeping entry. It turned out that "failure to swipe" was only the superficial manifestation of the underlying problem, which was that our buses were frequently late.

For people well skilled in problem identification and resolution, it is straightforward to examine problems for the root cause or the initiating event; for people without that skill, however, the most likely result is that improvement efforts will be directed toward the problem as it is experienced rather than the root cause of the problem. In this way, people work diligently on the symptoms of the problem, but that work does not yield improvement. Root cause analysis is not a natural skill and is not one that people frequently learn in their ordinary lives outside manufacturing. Therefore, if we want people to practice serious and productive industrial improvement, root cause analysis is one of several new problem-solving skills that management will need to provide.

The second classic manifestation of the difference between knowing about a problem and being able to solve it is that people who operate our plants often do not have the technical skill to solve the problems they experience. We have already discussed the issues of flexibility that arise when multiple products share the same equipment and the inventory that results from accommodating inflexible operations. The long experience of using the economic order quantity (EOQ) methodology clearly demonstrates that, for decades, managers recognized and attempted unsuccessfully to resolve the problem of inflexibility; however, they were not able to do more than accommodate the fact that the problem existed. The effective resolution finally came when the new tools of SMED (single minute exchange of dies) and FSVV (fixed sequence variable volume) were developed.

As discussed in Chapters 3 and 4, these and the other lean tools are great for use at the frontline. However, if we want people to use these tools for improvement, we need to provide them with the information they require to do so successfully. If managers and engineers were unable to figure out the problems with inflexible equipment before the solution was well documented, then it is likely that other people cannot do it without the documentation either.

Of course, many problems at the frontline can be recognized and resolved immediately using existing skills. Those improvements are often the basis for the first autonomous actions taken. However, we do not want random improvements or unfocused actions just because those actions are possible with preexisting skills. We want people to identify and resolve specific goal-oriented improvement opportunities that align with and contribute to the business strategies that we set out. Often, that requires that we provide people with new skills appropriate to the specific tasks and improvements that have aggregate value to the enterprise.

Liquid processors are typically quite good at delivering training, so the principal issue is to determine what training to give and to whom to give it. In precisely the same way that goals are both engaging and limiting, so are skills. Therefore, it is important to define specifically for each team the skills that it will use and the competence required before it can autonomously use those new skills to make the improvements described in its goals.

For some teams, the appropriate skills might be root cause analysis or some form of statistical training to allow them to analyze and improve basic operations. For other teams, the new skills might be SMED or autonomous maintenance to allow them to contribute increased flexibility or reliability. Other teams might start with mistake proofing, which will enable them to increase the success of their routine work. All the teams are likely to require some new skills, but probably no team requires all the new skills. A critical element of success in delivering new skills is for management to recognize and respond to those differences so that the needed skills are delivered without wasting people's time on training that they do not need or cannot use.

There are several important aspects to delivering new skills successfully. First, if you are asking your teams for very specific improvements, then you need to give them the skills that will enable those specific improvements. Never assume that your teams already have skills. It is all right—in fact, it is a good idea—to assess skills and avoid teaching skills that already exist. However, do not ask people to take on new assignments until you are sure that they have the skills to do that work appropriately and successfully.

Second, do not teach people skills that they will not use immediately. Competency improvement needs to be associated closely with the tasks that require the new competence. The most common error that managers make in delivering new skills is to require all people to receive training in a way that is convenient to the trainers, but results in training that the trainees will not be required or able to use for some time. When people learn skills that they will not use or will not use promptly enough to remember when the time for using them comes, both the training and the time spent in training are lost. Generally, this happens when a manager decides that a new technology is such a good thing that the training immediately becomes mandatory for everyone. Classes are scheduled and people are trained, but nothing of value happens as a result.

The third aspect of successfully providing training is that the new skills define what is permissible for autonomous action. As you commence autonomous actions, teams will be authorized to make improvements using the skills that you know they possess or the new skills that you provide. Equally importantly, they will not be authorized to make autonomous improvements using skills where they have not yet demonstrated existing competence or acquired the new competence required to succeed.

This does not mean that teams that have not been trained in a particular skill cannot undertake improvements that require that skill. It does mean that they cannot do it *alone*. They need to get help as they use that skill, or they need to get someone to do the new portion for them.

Key idea: Goals define what the team should do as well as what the team is not authorized to do. The same is true of skills. When a team is given a skill that corresponds to its goals, its members are expected to use it. When they do not have a particular skill, they cannot use that technology without help. The goals and skills are the foundation for ensuring that people do only the right things in only the right ways.

Time to Make Improvements

At work, management is the absolute owner of time. Periodically and informally, people may find a little time here and there to learn something new and to make a few modest improvements to their work, but improvement at a world-class pace cannot be accomplished informally. If we want to be certain that teams *do* routinely practice improvement, we need to provide them with some designated time when they *can* routinely practice it. This is normally easier in process plants than in mechanical plants. Quite often, our plants can run very well for a half hour or so with no action required from most of our operators. That is normally plenty of time to hold a satisfactory frontline improvement team meeting.

In most situations, it is best to plan to use each of the following three ways to provide teams with time to improve their work and meet their goals:

1. *Informal time* that operators create for themselves: Team members can use this time to conduct informal reviews of the quality stations and to talk among themselves about the improvements in progress and especially about new improvements that have been proposed but not yet been selected for action.
2. *Team meeting time* provided or scheduled by management: Quality station teams can use this time to meet to review progress, allocate resources and assignments, and promote new ideas into action. In most plants, this requires approximately 30 minutes a week, which is sufficient for team members to meet as an improvement team. They may also meet separately as a team for other purposes, such as toolbox safety reviews.
3. *Normal management control of time:* Management can provide front-line teams with specific time in which the improvement work of the team is included in the individual task assignments of technicians just as we schedule the other work of the technicians. When the team assigns an individual responsibility to execute a team task, management must provide the time for that work to be done.

The bottom line here is that in order to practice improvement effectively in addition to their normal work, the teams need time in quantities or in cohesive blocks that only management can provide. All team members need a modest amount of time to meet together as a team; consistent with the improvement work currently in progress, individuals on the team will need specific allocations of time as required to execute the improvement work assigned to them by the team. A block of time in which to meet and designated time as needed for project execution are resources that only management can provide. Without that time, there will be no significant improvement.

Access to the Resources That Cause Change

In industry, some things can be done essentially without cost or solely within the time and natural resources of the team. This is especially true if the teams have discretionary maintenance or operating budgets. A classic definition of small event improvement is the capability to improve with only the natural resources of the team. However, many opportunities for improvement arise where the team will not have natural access to the resources required.

FIGURE 10.1
Pipeline with clamped joints.

Case Study: Allocating Technical and Material Resources

In Suncor's extraction units, because the sand erodes the pipes, we spend a lot of time maintaining the pipelines that move the sand and water slurry away from the separation cells and return the now clean sand to the mine. Included in this maintenance, our practice is to rotate the individual pipe spools three times to reposition the zones of high erosion caused by the flow patterns that develop within the pipe. In that way, we substantially extend the useful life of the pipe before it needs to be replaced. Because these are relatively low-pressure lines, one of our teams proposed changing from the traditional welded seams used to join the pipes to Victaulic[1] couplings (shown in Figure 10.1) so that, when we rotate the pipes, they do not have to be cut and rewelded. The teams only need to open the clamps at each end of the spool, rotate the pipe, and retighten the clamps.

This conversion from one type of piping system to another represented a major improvement that the team could implement, but required resources beyond those natural to the team. Among other things, this autonomous team was not authorized to change the design or configuration of the pipelines without approval from a registered engineer. Further, the team naturally possessed parts, materials, and equipment appropriate to rotating pipes with welded seams, but not the materials and equipment appropriate for the proposed new system. This change was a great improvement that the team conceived and executed. However, between the concept and the execution, they needed access to some technical and material resources that were not naturally part of their complement.

If possible, it is normally good practice to allocate a small budget for this sort of thing so that it is within the team's control. This gives the teams greater autonomy when they are faced with a project that exceeds the preexisting resources natural to their prior limited scope of work. If they are "entitled" to a certain amount of engineering support or authorized to make a certain amount of unplanned spending decisions, teams often find a way to do amazing things with small budgets.

Conversely, when teams do not have any budget for this sort of thing and must ask for permission for each nonroutine project, they often bog down and quickly fail to do anything of interest. Teams that do not have access to the resources that will cause improvement can delegate their problem upward and assign responsibility for inaction to supervision. At that point, autonomous work has ceased. Particularly at the beginning of this effort, there is a very fine line between empowering people and discouraging them. Allocating a modest amount of resources to the control of the teams has great value to ensure that the teams are truly empowered to cause improvement.

Framework for Action

We discussed the framework for action, or quality station, as part of our discussion of policy deployment. I mention it again because, in addition to forming an integral part of the policy deployment process, the quality station is an integral element of engaging the people who will make improvements autonomously.

The quality station makes the process and outcome of autonomous improvement visually apparent. Here, the team demonstrates that, as the tactical execution of its goals evolves with time, it continues to maintain the original strategic alignment. The team also demonstrates that it is making significant progress. Of greatest importance, by making visible the goals and tactics, the completed work, the work in progress, and the planned work, the quality station allows managers and engineers to maintain unobtrusive oversight of the changes as part of a meaningful management of change effort.

People can do many things autonomously within a carefully structured system that defines what they can and cannot do and the ways in which they can and cannot do them. However, you always need to ensure that they never inadvertently exceed those limits to do something they should not do or do something in a way that they should not. The quality station facilitates that preventive caution in a way that no other system could. It allows managers and engineers to maintain oversight without intervening in the work of the team unless required. When intervention is needed, it provides an opportunity for engineers to intervene sufficiently early in the process so that the team efforts are not disrupted by the intervention.

ENGAGE FRONTLINE TEAMS

Normally, the frontline teams of your business are ready and willing to engage with you in improving the enterprise. They will usually have some ideas about what should change and the type of change required. Unfortunately, as you begin to engage with them, many of the team members' spontaneous ideas will not be related to the focused improvement of the business. As you engage the team, you are going to have to explain that you only want certain changes because world-class performance is achieved by focusing everyone on a few areas of real importance. Even the very best businesses cannot successfully change everything at once at a rapid pace. Changes—even improvements—that are not part of the focused system of improvement designed and deployed by management often distract from rather than help with achieving the greater goal.

Changing a few important things in significant areas and to a significant degree, as well as making those changes very rapidly, is the source of world-class performance. It may well be that many other things can be improved, but the only improvements that are authorized for autonomous action are the changes that align with the goals of the business. As always, management might decide to do other things and to engage the teams in them, or the teams may seek and obtain approval to do other things, but the team cannot initiate those other things autonomously.

The best way to initiate frontline improvement is through examples or pilots of the intended practice. As teams complete their goal translation activities, managers should work with them to select one goal-aligned tactic that is important to the business and interesting to the team. To be certain that the teams succeed, management should provide them with a lot of support as they undertake this initial project. As with autonomous maintenance, start deliberately and proceed in small steps that allow the team activities to grow in scope and in speed. The pilot projects do not need to be autonomous. They just need to be good representations of the sorts of things that the team should continue to do en route to becoming autonomous. Initially, the single critical issue is that these pilot projects should be both successful and exemplary.

When Exxon Baytown achieved 40 implemented improvements per person per year at year-end 1997, it was our sixth year in the business of engaging frontline teams. At the end of our first year, we enjoyed a rate of only eight improvements per person each year. That was quite a change from our prior rate, which could not have been much different from the North American average of 0.028 improvements, but it was not where we would be 6 years later. Start at a pace that you can sustain, proceed in a deliberate manner, and let the improvements make you better.

What to Do When Teams Do Not Engage

Teams at the frontline are normally ready to engage with you. The five objective elements of engagement followed by a few structured pilot projects and strong early support are normally enough to get them started in a productive way. After a good start, the interaction with the teams can quickly become a matter of routine management review and unobtrusive oversight.

Occasionally, some teams do not get started and require still more help. When a team appears to have all the elements objectively required to initiate autonomous improvement and still does not engage, you need to find out why its members are not engaged and work with the team to get them going. The first step in providing assistance is to review the five objective elements to be certain that your understanding of the situation is correct. The team may not sufficiently possess the five objective elements needed to engage in improvement:

1. *Goals:* A very common mistake that prevents teams from getting off to a good start is that they believe they have translated their goals, but they really do not yet understand the goals in a way that will enable them to describe tactics to implement their goals.
2. *Skills:* Some teams do not get off to a good start because they do not have the skills needed to diagnose and resolve the problems that are under their control and do not have appropriate access to someone who can help them technically with problem resolution.
3. *Time:* Some supervisors like to supervise, so they do not organize and provide time for the teams to act autonomously as problem-solving or improvement teams or do not allocate planned time to implement the work of the team.

4. *Resources:* Some managers and supervisors like to retain personal control of the resources of the business; therefore, you will find teams that have not been allocated any resources and thus cannot do anything autonomously.
5. *Framework for action:* Many team leaders do not spend the initial time needed to prepare their quality station in sufficient detail so that it supports autonomous work; as a result, it does not. The quality station appears to be a new element of "extra" work, but it is actually a necessary element of organizing and progressing autonomous work.

Key idea: When teams do not respond appropriately after you believe they have received the five objective elements of engagement, the first step is to check again to be certain that the five elements have been received in an effective way. Far too often, the reality is different from the report. *Some supervisors simply do not like to give up any form of control or do not know how to do so, with the result that their teams never quite get enough autonomy to succeed.*

Refresh the Understanding of Small Event Improvement

In the early days, it is also quite common to find teams that believe they require far more resources than they have been allocated. Although it is generally true that teams require some new resources before they will or can commence autonomous improvement, it is also true that they generally do not require many new resources. I am continuously amazed at how much the frontline teams are able to do with the resources normally within their control. By a wide margin, the best frontline improvements are improvements that teams can execute with only modest consumption of resources. *The concept of creativity before capital is especially visible at the frontline.*

When teams stall in their improvement efforts for lack of resources, it is often a result of following the wrong improvement model. We want the frontline teams to solve a new class of problems in a new way, but often the only model of improvement that teams possess as they start this process is the "engineer and manager" model of big event improvement. Engineers and managers resolve problems by making major changes or implementing a capital project. Often, engineers and managers solve problems by identifying them and assigning the resolution to others. None of those approaches is appropriate for frontline teams.

When you encounter this situation, carefully lead the team through another pilot project or two that clearly demonstrate their ability and your objective to have them implement improvement within their naturally occurring resources. Frontline teams always have abundant opportunities to do that, but sometimes they need help recognizing those opportunities because they look so different from the improvement opportunities they are accustomed to seeing engineers and managers address.

THE SUBJECTIVE ELEMENTS OF ENGAGEMENT

All that said, some teams do indeed have every objective thing they might need to get started and yet they still do not demonstrate any improvement. In those situations, it is likely that one or both of the subjective elements of engagement are disrupting the team's efforts. In that situation, it is likely that the team does not trust management and/or one or more members of the team are disrupting the team's social network. Both subjective elements of engagement are important and both appear at about the same frequency in industry, although one is normally more common than the other in any individual plant.

True autonomous improvement is a very creative activity and, for this reason, it is normally a voluntary activity. It is difficult to mandate creativity. Even though it is important to be very strict about not allowing teams to opt out and either refuse or fail to participate, it is also clear that autonomous improvement cannot be sustained successfully as a compulsory activity. When teams do not make progress, management needs to take action that will get them voluntarily back into the game—generally by discovering and resolving the reasons for their lack of engagement. You cannot simply require teams to become engaged by management fiat because, if they have everything that they need and are still not participating, there always is a strong and specific reason for it and they will not sustainably engage until you find that reason and correct it.

Generally, the cause for lack of engagement is one or both of the two subjective characteristics introduced before, which arise when the team members do not trust management or one or more members disrupt them and prevent them from effectively acting as a team. In the fullness of time, as you become good at managing small teams at the frontline, you can work proactively to develop trust and to form strong teams in very effective ways. At the beginning, though, the most common manifestation of these characteristics is that they disrupt the effort. Your response will likely be reactive, rather than proactive.

Lack of Trust in Management

Your frontline people will not give you additional improvements if they do not like what you have done with prior improvements—whether or not they were the source of those prior improvements. If management has an established practice of using improvements as an opportunity to reduce the workforce, then you have a long way to go before you can get the workforce to help you improve further. Indeed, any action that management takes that appears to be unfair has the potential to disrupt autonomous improvement. No action appears to be more unfair than reducing the workforce that has produced the improvement.

Fortunately, in the process industries, capital and material efficiency or even product quality improvements are generally of higher value than labor efficiencies. A companion concern, though, is that, for many people at the frontline, autonomous improvement, autonomous maintenance, and other new forms of activity are easily recognized as new work that they are being asked to do in addition to their current work. Even if it is interesting work that they enjoy doing, in terms of perceived fairness, adding new work to the current staff is thought of in much the same way as reducing the workforce.

There are, however, three good approaches to addressing the subjective issue of trust somewhat objectively. First, you can return the team's focus to the goals that they have adopted for themselves. Even when cost improvement is a major attribute of the original goal set of the corporation, very few teams adopt a translated version of the goal that contains elements of labor reduction as part of their action plans. Teams will focus naturally on improvements in material, capital, and quality. Remind them of that and set them off to improve the goal-focused objectives they have adopted that are unrelated to labor efficiency. Once you have them started, you and they will have the opportunity to demonstrate to each other what a good thing autonomous action can be.

It is also possible to offer your people a "guarantee" that they will not lose their employment *because of the improvements.* They may still forfeit their employment for bad behavior or for other reasons, including substantial changes in the market or in the business, but they will not lose their jobs as a direct result of the process of autonomous improvement. Typically, the factor that makes this guarantee possible is the presence of contractors, who can be replaced with employees, or purchased services that can be brought in house to provide additional work for your employees. I have made this guarantee both at Exxon and at Suncor with good results.

Case Study: Promises to Keep

Recently, Suncor initiated a major redesign of the organizations for maintenance, support services, engineering, and sustaining projects. The outcome of that effort is an estimated savings of more than $100 million per year. As we launched that initiative, I made the specific commitment that "no Suncor employee will lose his or her employment as a result of this change."

I made an identical commitment in 1991 as we initiated the improvements at Exxon Baytown. At Exxon, this meant that process technicians sometimes mowed the grass or were on loan to the United Way. Similarly, at Suncor, some people are working in jobs that are not precisely what they are accustomed to doing. In addition, in both companies, many outside contractors left and a great deal of previously purchased work came back in house. However, no employees at either Exxon or Suncor lost their employment *as a result of the improvements.* At the end of the day, at both Exxon and Suncor, the staff believed that the promise I made them was true and we all worked hard to demonstrate that it was. As a result, we made our businesses much more successful.

Another way to instill trust is to let the teams that do produce labor improvements keep the extra time gained from the improvements and use that time to further their autonomous efforts—that is, to use the time for additional autonomous maintenance or for another autonomous improvement. Because you will be using the same people for essentially the same basic work, this approach has no incremental cost. It has the added value of capturing the labor improvements that the team produces in a way other than reducing labor that nonetheless delivers tangible benefit to the business. In fact, teams that have already created some productivity improvement now have still more time to create more improvement. At Exxon, we improved productivity partially by reducing our contractor workforce.

We did not reduce our Exxon employees. We improved productivity of the Exxon workforce by greatly increasing our production volumes through their efforts.

Although this practice of allowing teams to keep their labor savings was originally conceived as a way for teams to earn their way out of the perceived extra work of autonomous improvement or to ensure the teams that the workforce would not be reduced by the improvement process, in practice it has had far more value than that. For many frontline people, autonomous maintenance and autonomous improvement are enjoyable and a nice change of pace, even if restricted to be highly focused on their formal goals. Allowing these frontline personnel to earn their way to even greater autonomy has become a real incentive to enhance the pace of improvement.

Key idea: As you consider these objective resolutions for teams that are not progressing, it is important to recall that the real issue remains the subjective concern of personal trust. In all that you do, you always need to demonstrate that management will treat people properly in the new environment. Even when engagement is something people truly want, it is a big change, and your employees will watch carefully to see what you do with it as it matures.

Disruption by Team Members

Teams that never truly begin to work together as teams will never produce successful autonomous improvement. Frontline improvement is especially sensitive to teamwork, which is one reason why we always talk about teams at the frontline rather than individuals. Therefore, when someone disrupts the team, it is possible that the team will not commence autonomous improvement or that its members will stop it if it has begun. The potential for disruption of team efforts is something of which management needs to be aware throughout the life of each team because it can occur at any time.

Fortunately, when disruption occurs, it is promptly visible to the managers who are monitoring the several quality stations because it will be apparent that the improvement effort has not started, has stopped, or is greatly diminished after an otherwise good start. The disruption of teamwork by an individual team member can be intentional or unintentional.

The key issue is that teams cannot possibly do good work if one or more of their members are disrupting the team's ability to function as a team; therefore, management must intervene to get the team back on track.

Key idea: It is important to be alert to the possibility that the person disrupting the team is the team leader or the second-line supervisor (the supervisor of the frontline supervisor). In some situations, the leader is being unintentionally disruptive; in other situations, the leader overtly or covertly does not want the new autonomy to succeed and is being intentionally disruptive.

Intentional Disruption

Intentional disruptions can arise from two different sources: (1) individuals or groups who do not want autonomous action to succeed, and (2) individuals who are naturally disruptive. The first category of intentionally disruptive people includes leaders who are concerned that the new style of leadership is one at which they will not succeed or in which they will be giving up the authority that constitutes their personal identity or is the culmination of years of effort. This category of potential disruptors also includes groups such as unions and union leaders who fear that they will not be as successful in the new environment as they had been or that the new environment will take away the organizational or personal status that they have developed over a long period.

In some cases, they are correct. There are indeed people who can supervise effectively but who cannot lead an engaged workforce with any amount of autonomy. Such people will need to find a new role as the organization evolves into a new way of working. Similarly, the role of unions will indeed transform as the relationship between the company and its employees matures. As you manage the transformation of your business, be aware of these possibilities. You may find that you need to reassign supervisors who cannot adapt and that you need to tread carefully around the union until it has found its new identity and role. Unions certainly continue to have a very important role in an environment of autonomous improvement, but it is somewhat different. I have often found that the best union leaders support this effort, but that is not always the case. Before commencing serious engagement at the frontline, it is always useful to assess and engage your union leaders so that this work is focused on improvement, rather than confrontation.

However, although some few supervisors will need to find new roles and you will have to work closely with the union leadership as they adapt, for most existing leaders and for all unions, it is possible to be fully successful in the world of autonomous improvement. Some time and attention are simply required to allow them to make the transition. This is another reason to start at a level of autonomy that you are certain you can sustain and then proceed deliberately. You need to recognize from the beginning that there are likely to be issues of this sort and reserve some management capability and capacity to accommodate the issues as they arise.

The second form of intentional disruption comes from a group of individuals that exists in every plant. They believe their mission is to be entertaining or disruptive, perhaps to add variety to the work lives of their colleagues or simply to entertain themselves. I often call them the "bad boys" or the "class clowns" of industry.

In traditional plants, the bad boys are always carefully on the right side of discipline but always very close to the edge. They are disruptive but not so disruptive that they get in trouble for it. In times of transition, such as during the initial days of autonomous improvement, they often step over the edge to test the waters, and they are always actively watching for any chink in management's resolve to implement new practices. A period of change is always a very attractive and active time for people who enjoy disrupting the work of a manufacturing plant.

In a plant with a traditional leadership style of strictly supervised work, it is the exclusive role of management to deal with this behavior, and the work-related impact of the behavior is solely between the individual and management. Unless the individual is personally offensive, these antics have very little social impact on the lives or performance of others.

As the plant's culture becomes more engaged and more autonomously team based, that social relationship at the frontline changes. There is always a social impact when people are trying to work as a team and a disruptive individual interferes with that work. In such an environment, disruptive behavior has a clear effect on the work of other individuals. In an engaged or autonomous workplace, management *cannot* tolerate any behavior that disrupts the performance of the team, even if it is the sort of behavior that might not have attracted much attention in the past.

This is a time to be very serious about the standards of conduct at work. Most people who are intentionally disruptive promptly get the message when confronted by a deadly serious manager. Some few are not able to modify their behavior and must be reassigned or even leave the company, but most of them will tone it down to an acceptable level if required to do so. Some, whose disruptive behavior was a response to the boredom of factory work, even become the best contributors because, in the new culture, they are no longer bored.

Oddly enough, intentional disruption is the easiest form of disruption to manage because it is the sort of thing that managers know how to overcome. They can do it in a very traditional manner, even in a new work environment.

Unintentional Disruption

Without doubt, unintentional disruption is the most difficult form of disruption for the teams attempting to engage in autonomous work. In this situation, good, honest, hardworking people want to come together and work as a team, but for some reason they fail. Without help, they will constantly be unable to get past whatever has come between them. Resolving situations of this sort is difficult for managers because it is often more like family counseling than management.

Unintentional disruptions are always interpersonal or social in nature. One person does less than the others think is appropriate. One person has body odor. One person talks too much. One person does not talk at all. One person behaves in a way that the others cannot understand. One person is habitually late. One person is critical of teammates. These and countless other things can disrupt a team and management needs to intervene to set it right. Teams with serious social issues are rarely able to resolve their problems alone.

As they intervene, managers often find themselves immersed in issues of human diversity. A black woman in a group of white men drops out of the effort. A Vietnamese woman is so polite that she never speaks while others are speaking, with the result that she never speaks at all. It becomes known that one team member lives an "alternative lifestyle" to which other members object overtly or privately. The possibilities are truly without limit. The critical impact on the business, though, is always the same. Unless management intervenes when a team becomes the industrial equivalent of a dysfunctional family, that team will not successfully contribute to an improvement effort.

Never assume that this cannot happen in your plant because it surely will. These are serious interpersonal issues, but the vast majority of teams can recover if you catch the issue before it becomes ingrained in the way members interact at a personal level. This means you *must* intervene as soon as it becomes apparent that a team is not participating in improvement in the same way as other teams.

The biggest problem is that, although managers are already good at dealing with the behavior of intentional disruptors, very few of us are prepared to deal with social issues. Because these social issues are always personal issues, when an intervention occurs, it must always be well done to avoid making the situation worse. For this reason, you will need to identify or develop some standby capability to make social interventions when they are needed within your organization. Often your business has this sort of social capability, but traditionally it probably was not deployed to team building at this level because the need for it never before existed.

Key idea: In childhood development, an early social stage is described as "parallel play." In that stage, children are physically close but their activities are independent of one another. For example, two girls might be sitting together, but each is independently playing with her own doll. That is a good description of the social state in most traditional manufacturing plants. Workers are physically close, but each has separate activities that are generally independent of the activities of others. At that level of social engagement, there is not much opportunity for interpersonal issues to disrupt the work of the team. As you engage frontline teams to conceive, plan, and execute their improvement work autonomously, the degree of social interaction increases far beyond anything that you will previously have experienced at work. You will need to be prepared to manage that as it happens.

INDUSTRIAL CULTURE

An engaged workforce is more like a cultural experience than a traditional business experience. The culture that you create for your business will be unique. It will be an industrial culture that reflects the needs and values of your business as well as the personal and social cultures that each member of your staff brings to work. Although you will necessarily create and manage this new industrial culture at a detailed level, you should keep your sights a little higher than normal and pay attention to the entirety of what you are creating. We said earlier that the purpose of deploying goals is not to possess goals but rather to enable people to implement those goals. The same is true as you create a new culture of engaged people at the frontline of your business. Your intent is not to have a new culture but rather to have a culture that enables people to practice autonomous improvement in small teams.[2]

NOTES

1. A trademark of the Victaulic Company.
2. Whole books have been devoted to the topics covered in this chapter, including mine: *A Culture of Rapid Improvement: Creating and Sustaining an Engaged Workforce* (New York: CRC Press, 2008).

11

People Development

INTRODUCTION

The last *cultural enabler* of the Shingo Prize is *people development.* As with leadership and ethics, this is potentially a huge topic. There are many approaches to developing people as well as many good reasons to do so. However, for our purpose, the primary focus is on developing people who are sufficiently competent to engage in the implementation and successful practice of lean process manufacturing.

For lean, statistical process control (SPC), single minute exchange of dies (SMED), and all other forms of autonomous improvement, people at the frontline play a critical role. To do that successfully, they often need enhanced competence because they will be undertaking activities that were not previously available to them. Thus far, we have generally focused on creating that new capability at the frontline, and we have assumed that the contributions of engineers and managers are an existing or fundamental capability of process manufacturing. This chapter will be different.

Development of competent people is an area where a substantial benefit is to be derived from paying closer attention to people *throughout* the business, including managers, engineers, and other professionals. As with lean manufacturing, a significant step on the path toward capturing this benefit is to adopt a new way of thinking about individual competence and the manner in which individual competence affects the performance of the business. Lean concepts and values created a new manufacturing paradigm; similarly, the new paradigm of industrial competence includes both adopting new practices and abandoning some misconceptions.

In this chapter, we focus primarily on the relatively new practice of specifically developing people in certain "critical positions" to have high competence. During Exxon Baytown's implementation of lean, this practice attained a new state of maturity and, as a result, it is commonly associated with lean practice in the liquid industries. Indeed, development of a few highly competent people for a few critical roles is probably another universally applicable lean technique that is uniquely more valuable to process manufacturers—once again, because of our greater reliance on technology and capital equipment, both of which demand more competence than does conducting or leading direct labor activities.

Note: In this chapter, the skills we have been discussing elsewhere that are needed to conduct frontline improvement are included among the elements of "basic competence" for operators at the frontline. This change is not intended to diminish the importance of those skills suddenly or to assume that they already exist; the purpose is to establish the intended baseline of competence that will allow us to distinguish the *additional* contribution that a few highly competent people in a few critical positions can make.

IMPACT OF COMPETENT PEOPLE ON ORGANIZATIONAL PERFORMANCE

A demonstrably direct relationship exists between the competence of the people in an organization and the performance of the business they operate. As the Exxon Baytown team undertook a formal assessment of this relationship, we were fortunate to have access to Exxon's global resources. With the help of our colleagues around the world, we were able to assess many operating organizations, in many places, that were responsible for a wide portfolio of processes in various states of maturity and with varying performance. Since retiring from Exxon, I continue to gather information from still more organizations that have added a real diversity to the data.

The important *new* element of the relationship between competence and performance is that there are two substantially different aspects of industrial competence. They must be managed independently, but together they substantially determine organizational performance. *Basic competence* is required in nearly all people before an organization can achieve stable and safe performance and industry-average business results. *High competence* is required of a few people in critical positions before an organization can demonstrate a pace of improvement and business results that far exceed industry standards.

Our data indicate that the mix of high competence and basic competence exists in many combinations. However, it is clear from the data that two very specific thresholds define the relationship between individual competence and organizational performance:

1. *Either* ubiquitous basic competence *or* highly competent people in critical positions are required for an organization to achieve stable, safe, average performance.
2. *Both* ubiquitous basic competence *and* highly competent people in critical positions are required for an organization to achieve "best in class" performance.

We will discuss both aspects of competence, with an emphasis on the creation of the highly competent people who will allow your organization to achieve a new level of performance.

Competence Defined

Because our intent is to develop people to have sufficient competence to practice lean manufacturing successfully, the applicable definition of competence is both immediate and practical. For this purpose, competence is an assessment of an individual's *demonstrated performance in the specific role* that he or she currently occupies.

Prior excellent performance in other roles and potential excellent performance in future roles are both important for many purposes, but not for this. If a person's excellent abilities do not translate to real-time excellent performance in the person's current role, those abilities do not contribute to current organizational performance and do not count in this analysis. A person who possesses great capabilities, but who currently functions in an average way, has no more value to the business than any other average performer.

Key idea: This assessment of competence is limited to the demonstrated performance of one specific person in one specific role.

TABLE 11.1

Competence Matrix

Competence level I	A person is able to conduct only the basic elements of the task and who requires supervision to achieve that performance. This is the level of competence normally associated with a poor performer or someone who is new to the job.
Competence level II	A person is able to conduct all the basic elements of the task with no supervision. Such a person may be able to do more advanced elements with some supervision and ought to be able to create and implement small event improvement associated with the basic elements of the work. This is the level of competence and performance that should be normal to most of the workforce.
Competence level III	A person is able to perform all of the task elements, both basic and advanced, with no supervision. Such a person ought to initiate improvement to the basic elements of the task frequently and regularly initiate improvements to the more advanced elements of the work. This person also provides task help (normally mentoring, not supervision) that will enable others to achieve competence level II.
Competence level IV	A person is a master of all the elements of the task and is competent enough to improve both the basic and advanced elements of the task frequently and to create valuable extension to the work itself. This person provides task help (normally mentoring, not supervision) that will enable others to achieve competence level II or III.

Basic Competence

As either *Responsible Care*™* or *process safety management* documents, it is well recognized that an organization *cannot* succeed unless its people have the fundamental competence to conduct the business in a stable and safe manner. Based on the descriptions contained in Table 11.1, the responsible care guidelines require that essentially all the people in your organization should be at competence level II.

The Responsible Care™ focus on competence level II recognizes that in the process industries, an ever present potential exists for something of serious consequence to occur. In that environment, to prevent serious occurrences or to respond properly, people must generally have full capability to act independently and successfully in at least all basic elements of their work. Although the standard emphasizes matters of safety and environmental performance, the same attributes of personal competence apply equally to all other aspects of process operation. In aggregate, all

* Responsible Care is a trademark of the American Chemistry Council.

attributes of basic competence determine the business's fundamental performance. Certainly, frontline teams cannot practice lean manufacturing if they are struggling to conduct their routine work successfully.

At competence level II, most people in your plant must satisfactorily perform the essential elements of their work within their own capabilities. The tolerance provided by describing this requirement as "most" rather than "all" recognizes that some people will be new in their roles and some people will not routinely perform as expected. This situation of routinely employing some people of lesser competence is normal in industry and your process must be robust enough to accommodate it.

When most people in your plant routinely perform in a satisfactory way within their own capabilities, leaders have an opportunity to provide special attention to those who cannot or do not regularly meet that standard. With such a structure in place, successful execution of the many detailed aspects of operating a chemical plant can assuredly avoid the potential consequences of operating errors. Most people will have basic competence and those who do not will have a fully competent supervisor or mentor available to help meet operating expectations.

Case Study: Identifying the Inexperienced

At Suncor, we typically wear blue hard hats in our plants, but new employees *must* wear green hard hats until they have demonstrated basic competence at their tasks. This makes it visually apparent if someone who has not yet demonstrated basic competence is working alone.

Competence level II can be considered to be the threshold for ordinary performance within the liquid industries. There is nominal allowance for a few people who are new and for a few people who perform badly. However, unless 90% or more of the organization is at least at competence level II, the resulting performance is likely to be substandard.

Some organizations artificially emulate "ordinary" performance by diverting highly competent people, who otherwise represent the organization's improvement capability, to assist in sustaining the basic operation. In this way, a great many organizations miss the opportunity to improve. They accommodate lack of level II competence throughout the organization by requiring highly competent people to work routinely below their capabilities. This is often an effective means of facilitating a new plant, major expansion, or another event that requires the simultaneous hiring of many untrained people, but it is a great waste of talent if it is a routine practice to accommodate the lack of development of a stable workforce.

Key idea: Level II competence is the threshold to ensure that your operations will be routinely stable and safe.

Basic Competence Development

Competence is an indicator of the performance of one specific individual in one specific role. This implies that competence improvement is also a highly specific activity to develop the capabilities of a single person to match the requirements of his or her role. Fortunately, the roles of most people in any organization are sufficiently interchangeable that a satisfactory level of basic competence can be achieved with common development programs that can be easily shared among many people. Thus, although it is true that competence is measured by individual performance in specific roles, it is equally true that many of those roles are not unique and do not require substantial individual development.

The most common factor that gives rise to the lack of basic competence development is that routine jobs in many process plants are insufficiently documented. That is a real oversight. Pipe fitter apprentices receive 4 years of well-documented development, including classroom training and on-the-job experience. Many process operators often get little more than a few months of "stand by" development (As in, "John is a good operator. Go stand by him and see if you learn something."). The first task in developing people for lean process manufacturing is to document the basic requirements of our work thoroughly and ensure that everyone has at least that capability.

Key idea: *The basic competence of most people determines the ability of an organization to improve.* Organizations that possess basic competence will allow highly competent people to make special contributions. Organizations that lack basic competence divert highly competent people and prevent them from making special contributions.

Superior Performance

At Exxon Baytown, we undertook our assessment of competence in an attempt to find a management tool that would allow us to develop people who were more highly competent in a way that would directly improve performance. We strongly believed that the competence of individuals must relate to performance, but we all had long experience with industrial training efforts that routinely produced little or no impact on performance. We were looking for a new approach.

Our study of the relationship between individual competence and organizational performance clearly demonstrated that level II competence was necessary for routine operation and basic frontline improvement. We found that it was possible to determine objectively the required basic competence that would enable a business to perform according to industry standards, and it was easily possible to determine if that level of competence existed. When it did not, the plant routinely performed badly, or the plant routinely diverted a few highly competent people to accommodate the deficiency and, in that configuration, all aspects of routine improvement stopped. These were all easily observable objective attributes of plant performance. That direct relationship makes it possible to manage the basic competence of an organization in a way that facilitates predictably stable operations.

Our study also demonstrated that level II competence alone was insufficient to achieve the best possible pace of improvement. Just as level II competence is the threshold to achieving routinely stable and safe operations, level II competence in most people—plus level III competence for some people in certain positions—is the threshold to achieving world-class performance.

Key idea: In order for an organization to survive, most people must demonstrate at least basic competence. In order for your organization to move beyond ordinary results to become one of the best manufacturers in your industry, some people must be especially good at what they do and they must be positioned for their contribution to have a special impact. The people who have a surplus of competence have the ability to help others and the ability to recognize and implement improvements that are not obvious to others.

To our surprise, in Exxon's global assessment of the relationship between competence and performance, we initially found that superior competence is indeed required for above average performance, but *the precise correspondence between superior competence and superior performance was inconsistent.* Some organizations had many people with superior competence but achieved performance that was not meaningfully different from organizations with relatively few people of superior competence.

In other situations, we found that organizations with more people of superior competence achieved worse performance than organizations with fewer highly competent people. In our initial data, the correspondence between superior individual competence and superior organizational performance existed, but it was too weak to become a useful management tool for creating improved results.

Critical Positions

The breakthrough that allowed us to manage superior competence as a tool for reliably achieving superior performance was our discovery of critical positions. Our assessment demonstrated that, in some positions in organizations, competence has special leverage or influence on the performance of the entire organization. When an organization is stable, because most people are generally at level II competence, a few people of special competence (level III or level IV) *who occupy critical positions* enable the organization to achieve special performance in a predictable and repeatable way.

Key idea: A critical position is an organizational role where excellent performance within that role will have a positive effect on the business mission of the organization.

Once we understood the nature of critical positions, all of our assessment data from many different organizations became meaningfully aligned with our expectation that individual competence could be managed to make a direct contribution to organizational performance according to two simple rules of practice:

1. The extent to which an organization possesses people who generally demonstrate level II competence reliably and predictably determines the ability of that organization to perform according to industry norms.
2. *Equally,* the extent to which an organization fills *critical positions* with highly competent people (level III or level IV) reliably and

predictably determines the ability of that organization to *exceed* industry norms and to become truly excellent.

FINDING THE RIGHT MANAGEMENT TOOL

To create a new management tool, you need to establish a reliable and predictable relationship between the business attribute to be managed (in this case, competence) and the business outcome that results from that management (in this case, performance). With a good understanding of that relationship established for this purpose, we could create a new management tool or practice, confident that placing or developing highly competent people in critical positions would have a beneficial impact on performance.

A Quick Description of Our Analysis

After assessing more than 200 separate organizations, we created two rank-ordered sequential lists of results. On one list, we sequenced organizational performance from best to worst. On the other, we sequenced organizations from highest to lowest according to the extent to which their critical positions were occupied with highly competent people. After allowing a modest tolerance to accommodate for measurement error, the performance of more than 95% of the organizations aligned into the same sequence on both lists. Our analysis demonstrated in a statistically significant way that *the performance of organizations corresponds very predictably to the placement of people with special competence into positions of high leverage.*

Note: All of the data reported here were gathered from the process industries. I do not know whether industries that are heavily reliant on unskilled labor would experience the same result.

The Influence of Critical Positions on Improvement

Critical positions allow a highly competent person to leverage that competence into improved organizational performance in three different ways: *individual contribution*, *mentorship*, and *leadership*.

Note: Although leadership is one of the three areas of special contribution, one of the major misconceptions about competence—that all leadership roles are critical—must be abandoned because of this assessment. Our data show that this belief is not true or even close to true. By quite a wide margin, most of those in critical positions do not formally lead or manage others. Certainly, there are critical leadership roles, but most people in critical positions make a very *direct* contribution to the success of the business rather than the *indirect* contribution made by leaders.

Let us now look at each of the three categories of critical positions to understand how they influence organizational performance.

Individual Contributors

Positions that enable people to make special *individual contributions* typically are those where the organization has a clear need for real expertise in a particularly valuable aspect of the work. When that particularly valuable work is done exceedingly well, the business is demonstrably better for it. At Suncor, for example, we produce our own bitumen from multiple sources and we produce it in two very different ways (mining and in situ extraction). We also import bitumen from other sources produced by other companies. The properties of the bitumen from each source are influenced by the natural characteristics of the mine in which it is found as well as by the process used to produce it. From this diverse raw material, we produce raw bitumen, dilute bitumen, sweet synthetic crude oil, sour synthetic crude, regular diesel fuel, and low-sulfur diesel fuel.

We have many analytical capabilities, and many people contribute to the outcome; however, at the end of the day, a single individual who comprehends the entirety of our business optimizes the raw material that we acquire, the way we process it, and the products that we make. This is not a management position, but the success of this individual can influence our profit each day by 5% or more. Similarly, at Exxon's oil additives business, we had a position for a single individual who determined the blends of available materials that would serve our customers and optimize our profit.

There are many such positions throughout the liquids industry. Other examples of jobs in which individual contributions are paramount are analytical chemists, control console operators, maintenance planners, and reliability engineers. Although they are normally less dramatic in the extent of their impact than the operations planners described earlier, these positions and many others require the same characteristic: a single highly competent person who can make a uniquely valuable contribution to the performance of the business.

Case Study: Individual Impact on Performance

At Exxon Mobil, following the merger, I was helping one of our plant managers assess the cause of rapidly declining performance in a specialty chemical plant. During that assessment, it became obvious that the plant had previously enjoyed the services of two highly competent analytical chemists who supported operations. Immediately following the merger, both had retired. A person of modest competence replaced one; the other had not yet been replaced. These individual contributor positions were effectively invisible to the new management of this plant and, prior to the formal assessment of critical positions, management had no idea of the impact that losing these two individuals would have. Following the assessment, we rehired one of the retired chemists as a contractor and commenced serious development for the occupant of the other position. The result was a prompt and valuable improvement in performance returning to the standards previously achieved.

Subject Matter Experts or Mentors

Individuals also can make a special contribution through *mentoring* a field of technology or operations as well as all the individuals who practice that technology. At both Exxon and Suncor, we refer to such people as subject matter experts (SMEs). Normally, we have one SME for each important field of technical practice. This might be a chief metallurgist or a senior instrument technician. Similarly to the individual contributors, the people who are mentors generally fill roles that demand a specific technical competence.

In most cases, in addition to helping others, mentors use their special competence to produce a direct benefit. When a unique capability is needed, the SME can deliver it; however, unlike individual contributors, mentors do not always, or only, contribute directly to business performance. Instead, their role is often to help others use the technology that they mentor so that the entire group of technical practitioners becomes better able to make special contributions.

Mentors not only improve the quality of work of the other practitioners of the technology, but also improve the technology in ways that are of special value to the business. For example, as a consequence of the nature of our business, Suncor has developed metallurgical expertise in erosion and corrosion that is uniquely appropriate to our industry and uniquely valuable to our business performance. When a technology is important to the business result, an individual mentor for the practice and development of that technology can help everyone do it better.

Critical Position Thinking

We discussed earlier that "statistical thinking" can enable improvement without the need for statistical analysis. It is also possible to benefit from "critical position thinking." It is clear that if a technology is sufficiently important to your business, you will need an SME to mentor that technology and the people who practice it. The "critical position" concept can be applied to the recognition of subject matter experts without the rigor of a formal analysis. Most technical professions, including the engineering disciplines and the craft specialties, will benefit from the mentorship of a person who possesses truly superior skills and who can convey those skills to others. If a technology is important to your business, then it is likely that you should have someone on your staff with special skills in that technology. This person is not a technical manager, but rather is someone with superior technical skills who can improve the technology as applied to your business and can help other people practicing the same technology to do it better.

Case Study: The Benefits of a Super Craft Technician

In one chemical plant that continually experienced a series of small but disruptive equipment failures, we decided to create a "super craft technician" in each trade area. We needed to develop these people ourselves because no one in the plant with special skills existed in any craft role. Although this development took some time, the results were just what we intended them to be:

1. The presence of a recognized expert in each field allowed us to know precisely whom to assign to ensure that the most complex tasks were perfectly executed.
2. A recognized mentor allowed the other craft technicians to obtain advice and on-the-job development when they encountered situations requiring greater capability than they possessed.

3. With a mentor for each craft, all the craft technicians began to perform at a higher level and the mechanical problems quickly diminished.

Case Study: Supercompetent Engineer

In a more complex example of this practice, at Suncor, our large mobile equipment routinely experiences structural problems because of our severely cold Arctic winters. We hired an individual with combined technical capabilities in mobile equipment, structural engineering, and metallurgy. He made a big contribution but, more importantly, he served as a mentor to all the people who maintain our fleet. As a result, the winter of 2008–2009 was the first year in Suncor's 40-year history where we completed the winter season with zero structural failures in our heavy equipment.

Leaders

The third critical position is *leader.* This leadership role is not fungible (that is, indistinguishable from many others), and it would not include most of your managers and supervisors. This leader does not simply manage an arbitrary group of people within the plant. Rather, this type of leader *leads and mentors* a group that, together, performs a critical role. For example, if several individuals jointly did lube oil blending, then leading that group might be a critical role. This would be especially true if the leader had skills to improve the field of practice and to mentor the individual members of the team.

The distinguishing element of this role is not management or even leadership per se, but rather *leadership that,* in combination with special competence in a critical skill, *raises the performance of the organization.*

Identifying Critical Roles in Your Organization

With no more than the understanding that critical roles are positions where an individual of special competence can make a contribution of unique value to the business, it is normally possible to commence effective competency management promptly by identifying those positions and either developing or assigning highly competent people to fill those roles. Supported by descriptions and examples of the different types of critical roles, this process can be accomplished easily and quickly even if you have not previously done it.

Common Misconceptions

The greatest sources of problems in using this management practice are three pervasive misconceptions about competence and leadership. First among these is that many people want to describe all management roles as being critical to the business. The second is the failure to distinguish work that is necessary from work that is critical. Third, people are frequently confused by work that is important but fungible. The essential problem with all three of these misconceptions is that they result in identifying far too many positions as being critical. The intent of this exercise is to identify a focused few positions where we can effectively manage the competence of a few people in a way that will improve the performance of the organization. Let us take a minute to clear away those misconceptions.

Roles That Are Important but Fungible

One of the most common sources of confusion in managing critical positions is that a role can be important but, at the same time, fungible (or indistinguishable). People often think that if a role is important, then it must also be critical, but that is not true. The characteristic of fungibility is that there are many individual roles, each one indistinguishable from the others. The typical example used to illustrate this characteristic is grains of rice. It is practically impossible to distinguish one grain of rice from another and there is generally no value in doing so. In industry, it is common for many people to occupy fungible roles in a way that the aggregate outcome of their shared work is very important to the business. However, these roles are not critical positions because it is not possible for one individual of special competence in these roles to make a personal, positive impact on the business. *The test for criticality is whether a single individual has the ability, through the application of great skill, to do the work so well that the business outcome is improved.* That is very rarely the situation when many people do the same work.

When people attempt to describe the critical impact of important but fungible roles, it is usually a matter of negative, rather than positive, consequence. It is true that all the pipe fitters in your plant stopping work would have bad consequences, and it is also true that one individual pipe fitter can do the job so poorly that he alone produces bad consequences. Pipe fitting is clearly important work in a process plant and either of these possible outcomes would unquestionably be bad, but that does not make the role of pipe fitter critical. The distinction is that, in assessing the criticality of positions, you are searching for roles where it is possible to produce improvement through superior skills. The issue of avoiding detrimental performance resulting from poor skill or poor execution is normally an issue of level II competence or routine performance management, but not an issue of the criticality of the role.

Key idea: The concept of critical positions is defined in a specific way for this very specific purpose. Other definitions of "critical," including related words such as "important," "essential," or "vital," are irrelevant and often confusing in this context.

At both Exxon and Suncor, pipe fitters clearly have an important role, but these positions are not critical because we have several hundred pipe fitters. There is likely to be a role for a senior technician to mentor this technology and to assist the pipe fitters as a group, but it is not generally possible for any individual pipe fitter to do his or her work in a way that improves the performance of the business.

Roles That Are Necessary but Not Critical

The second source of confusion in the search for critical positions is roles where the function is necessary, but not critical. For example, it is necessary to pay the bills. Although there is often no issue of fungibility because only a few people are normally involved in activities such as paying bills, these positions are not critical because there is no possibility that anyone could pay bills in such a superior way that the business performance becomes demonstrably better as a consequence.

Once again, people who attempt to include necessary work in the determination of critical positions do so by describing the negative consequences of poor performance rather than improvement resulting from superior performance. Many business activities would produce negative consequences if the activity simply stopped or if it was performed very badly. Activities such as paying the bills, carrying out the trash, or distributing the mail fall within that category. Some of these roles are unique to the operation; for example, in northern Alberta, snow removal from our plant is necessary so that people can do their work without falling. Although these necessary roles can be done so poorly that there may be a negative impact, none of them can be done so well that it will have a positive impact on the business.

Leadership Positions That Are Important but Fungible

By a very wide margin, the most common mistake people make when assessing critical positions is to assume that most supervisory or managerial roles are critical. The work of leadership is always important and necessary, but it is not always critical. Our detailed assessment of many organizations demonstrated that, at the extreme upper end of the managerial spectrum, only three of the five most senior leaders of a business typically occupy critical roles. That ratio of critical positions decreases rapidly as the assessment moves to managerial and supervisory roles with less strategic content. A defining attribute of most critical positions is that individual excellence in performing the role can make a special contribution to the success of the business; most leadership positions do not objectively demonstrate that attribute.

Key idea: It is a good thing that not all leaders are in critical positions. If all leadership roles required individuals who possessed specific high competencies in their positions, we could not easily assign people to roles that are far removed from their prior experience. This would greatly diminish our ability to develop leaders who have broad exposure throughout the business.

DEVELOPING HIGHLY COMPETENT PEOPLE

There are two noticeably separate aspects to industrial competence. The first is to ensure that all people have basic competence; the second is to ensure that the people in critical positions have high competence. They are both important and together they are necessary for superior performance. However, they are successfully managed in very different ways.

When you are striving to establish level II competence broadly across the organization, the existence of fungible positions is a great benefit. Competence in all roles is always a matter of developing a single person to perform well at a specific task. Riggers do not get the same development that pipe fitters get. However, all riggers do get the same basic development to succeed as riggers and all pipe fitters get the same basic development to succeed as pipe fitters. Some basic development, such as the plant's general safety standards and practices, is the same for all people in all positions. In this way, level II development is normally standardized and shared. The same development program can be used for many people over a period of several years and relatively large groups of people can be successfully developed together.

The same is not true when a business is attempting to develop highly competent people who can make special contributions in critical positions. High competence is difficult to achieve. It normally requires a careful assessment of the existing capabilities of an individual and the characteristics required for success in the individual's unique role. As a result, in most cases, the development required for highly competent people in critical positions is an activity that is unique to each individual.

The most common problem of managing competence in critical positions is that many businesses have trouble making the transition from large-scale development of basic skills to specific development of special skills. This difficulty manifests itself when they attempt to develop people beyond basic competence using generic programs delivered to large groups, just as they do to achieve basic competence. Applying that technique to develop advanced skills is painfully slow, extremely difficult, and often too expensive to sustain. Furthermore, such efforts often do not create highly competent people with special skills.

The solution to the problem is to concentrate the development of special skills only on the few people who do or will occupy critical positions. It is genuinely not practical and, according to our study, not immediately valuable to attempt to produce high competence in all your people. However, it is both practical and valuable to develop high competence in the few people who will occupy your critical positions. By focusing your initial development of high competence on the few who are in critical roles, you can deploy the resources needed to achieve success. Developing level III and IV competence is very different from developing level II competence, but by starting with critical roles, it is possible for most organizations to make rapid progress.

Beginning the Process

The first step in using critical roles as a focus toward developing special competence is to identify which positions are the critical positions. Immediate opportunities to benefit from this understanding of critical roles are available even without a formal assessment, but the *full value* to be derived from this concept is only possible when a business undertakes the management of critical positions in a *formal* and *sustainable* manner. As the example of the analytical chemists demonstrates, having highly competent people at one point in time does not guarantee that you will always have such people unless you take specific action to manage that outcome.

We have talked about the possibility that an organization will determine that it has many critical positions simply because those positions are important, necessary, or leadership roles. At the extreme, organizations that make this mistake while attempting to manage critical roles identify so many critical positions that they promptly return to practicing broad-based development under a different name. The reason for carefully limiting the number of critical positions identified is that we do not simply want to possess a list of critical positions. We want to manage the competence of the people who occupy those critical positions, and we want to achieve the benefit to be derived from the special competence of a few people in highly leveraged roles promptly. This always requires a focused effort, which is not possible if the identification of critical roles is not closely managed.

Key idea: In my experience, the total number of critical roles in most organizations is between 8 and 12% of all *positions.* As is common in process plants with 24-hour continuous operation, more than one individual will occupy some of these roles. Nevertheless, the number of *people* in critical roles is normally no more than 15% of the total population because many critical positions are individually occupied or due to individual contributor roles such as subject matter experts.

Limiting the identification of critical positions to a few roles with highly leveraged influence enables you to initiate a "people development" effort at competence levels III and IV in a way that is sustainable and that will have an immediate and beneficial impact on your business.

As you assess roles to identify the critical positions, it is useful to document the characteristics that make each role critical because they also define the development needs required to fill that position successfully with a highly competent person. As Henry Ford observed, "No task is particularly difficult if you break it down into small pieces." The same is true of developing people to fill critical roles. Once you know the several characteristics that make a role critical, the path to developing people who have those characteristics is more straightforward than you might have imagined.

Developing highly competent people in a carefully structured way to match specific skills to the specific needs of a critical role is difficult; however, with focused efforts, you can successfully deliver very specific training to a few people. The principal errors to avoid are the dilutive impact of defining too many critical positions, the temptation to focus all training on leaders, and the wasteful effort of delivering *bulk training* in place of *focused development.*

Prompt Improvement

In most plants, you will find that you already have some highly competent people in critical positions. As with many other aspects of lean practice, the formal assessment of critical positions makes common and formal the valuable contributions previously made informally by a few people. It is not surprising to find that some people have spontaneously become highly competent.

Of real importance is the fact that, in most plants, you will find other people who could immediately demonstrate high competence in a critical position if they were assigned that role. As a result, in most plants, it is possible to reassign a few people quickly to roles that better utilize their existing capabilities. As a result of highly competent people who are already in critical positions and those who can promptly be assigned to critical positions, it is normal that, after a development effort focused on 5% or less of your workforce, you can quickly reach a situation in which you have at least one highly competent person for each critical role.

Key idea: It is common to find that the current supervisor of a critical position previously held that role and performed in a highly competent manner. This does not mean that the supervisory role is critical and you should not demote the supervisor. However, it does provide a clear vision of what a highly competent person looks like and a good opportunity to provide on-the-job development of the incumbent with the highly competent supervisor as a mentor.

SUSTAINING THE IMPROVEMENT

A vital aspect of critical position management that very few organizations do well today is managing these positions sustainably over time. When assessing an organization to identify critical positions, it is common to find critical positions that do not at the time have highly competent incumbents, although highly competent people once held those same positions. It is even common to find critical positions that are currently unoccupied. At Suncor, I found some critical positions that once made great contributions to organizational performance but that no longer existed on the organization charts.

There are always formal and well-developed practices to manage competence and succession for managerial positions. However, for individual contributors, including those who occupy critical positions, that development rarely happens in a sufficiently formal manner. Nevertheless, it is not a difficult task and it does not require new procedures. Simply apply the existing capability for management development and succession to this new population. It will likely involve some new people in the succession planning and development process. For example, the executive committee is responsible for succession planning for vice presidents, but it is maintenance managers who are responsible for succession planning for senior instrument technicians. Ensuring that important roles are always filled well is straightforward. Just remember to do it.

Key idea: When we originally assessed many Exxon organizations, we found that the correspondence between organizational performance and the proportion of highly competent people in critical positions extended beyond 100%. That is, when a plant had critical positions fully staffed with highly competent people, relative performance continued to improve as the plant added still more highly competent people who also focused on those critical roles. These additional highly competent people provided capable stand-ins during vacations and other absences and they provided prompt and fully qualified replacements when the incumbent moved on for any reason. Therefore, although you ought to initiate your use of this new management tool by focusing on the creation of a single highly competent person for each critical role, once that has been achieved, additional benefit is still possible through further focused development.

Managing general competence is demonstrably beneficial to ensure that all people achieve at least basic skills. Managing the competence of a few people in critical roles to achieve high skills has an added benefit. In most situations, it is possible to obtain this benefit promptly simply by recognizing the issues and engaging the organization to focus on the people and roles that have the most to offer. Ongoing formal management of competence and critical positions is necessary to sustain the gains and makes it possible to obtain even further gains in the future.

12

Leadership: Initiating and Sustaining Lean Operations

INTRODUCTION

Lean manufacturing is such a pervasive approach to operating and improving your business that it is often overwhelming for a leader to decide how to get started. All the lean elements and tools seem like good ideas (and they are), but how do you begin? In addition, most leaders who have been around for a while are as concerned about *sustaining* change once it has been implemented as they are about initiating it. In this final chapter, we will explore how you can initiate lean manufacturing and can sustain it over time. When we have finished, you will understand how you can adapt the practice to your own business and people.

Before beginning a detailed discussion of how to initiate this transformation, we need first to establish a shared understanding of what constitutes an organizational transformation. Also, we need to examine the distinct role that individual leadership plays in achieving that outcome.

TRANSFORMING AN ORGANIZATION AND SUSTAINING THE CHANGE

The reason to adopt lean manufacturing is to *transform* your business: to make it *substantially* better than it has been. The critical element of the lean transformation that most people do not adequately recognize is that a transformation of this magnitude is not a one-time event. You will not operate a traditional business one day and a lean business the next. The initial transformation to make lean practices common and stable within most businesses occurs over a period of approximately 2 years. Following that, the lean culture enables a long period of very rapid improvement as the new practices mature and as people become well accustomed to this engaging way to work.

This does not imply that you need to wait 2 years to experience the benefit—just the opposite. You should *immediately* experience the benefit of each new lean practice that you adopt in each place where it is adopted, and you should *continue* to experience the benefit at an accelerating pace throughout the implementation as the new practices expand and mature. Moreover, lean should greatly increase the agility with which your business responds successfully to changes, challenges, and opportunities that occur in the future. However, recognition of this extended path for the adoption of lean does imply that equally careful attention needs to be given both to initially implementing lean and to sustaining lean through the period of maturation and beyond.

Case Study: Continual Transformation

At Exxon Baytown, as we increased the capability and capacity of our plant, we initially began to produce more volume from existing assets. Next, we began to make better products, still at higher volumes. Then, we began to combine improved capacity and capability to make more difficult, higher value products—again, at higher volumes. Each time we achieved some new level of performance, the next opportunity for still better performance was both readily apparent and achievable. After an amazingly successful beginning that resulted in being designated one of America's 10 best plants after our first 2 years, for over 5 consecutive years, our pace of improvement accelerated each year.

When you truly transform your business, improvement is neither modest nor slow; rather, you should anticipate immediate benefits and you should experience the magnitude and speed of lean transformation accelerating with time. One of the most satisfying experiences of lean occurs as you and your team realize that you are continually doing things that you previously believed were not possible. As you develop confidence in your new capabilities, you will begin to plan for outcomes that are currently beyond your reach with the expectation that they will become possible as you continue on your trajectory of improvement.

Case Study: One Improvement Leads to Another

At Suncor, as we increased the capability and capacity of our extraction units, we developed sufficient new capacity (with no new investment) so that we were able to run extraction much more intensely. In that mode of operation, we reduced the average amount of sand entrained in the bitumen delivered to the upgrader by half, and we essentially eliminated episodes of process variation where the amount of entrained sand spiked to very high levels. With less sand, we experienced less erosion, and with less erosion, we experienced less downtime to repair the erosion. With less downtime, our upgrader produced more oil. Every improvement led directly to another.

With this improvement and many others, in a business accustomed to growing principally through capital investment, within 6 months we substantially increased the output of existing assets. Having achieved a result previously considered impossible, we changed our view of the future. Previously, in periods when access to capital was limited as it is today by low oil prices, we described that situation as one of "no growth." Today, we describe that situation as "organic growth," and our current plans call for continuously expanded capacity at a rate far exceeding rates normally achieved in the energy industry.

The essence of transformational leadership is a prompt and ongoing business outcome of significant magnitude. Strategically focused and tangible aspects of your business results will change in a very real way. The effect of this "real" transformation will be apparent: more production, better quality, improved customer service, new products of higher value, and so on. By continuously looking for the objective results of lean in your business performance, you can assure yourself that you are using lean to achieve the *substance* of becoming a better manufacturer and not simply the form of operating in a lean way.

Managers who lack a clear vision of the future to serve as a reference often introduce lean in a way that emulates what others have done, but without achieving a benefit to their business. Many of these "copy cat" initiatives are so superficial and inappropriate that they harm rather than benefit the business. Alternately, a transformational leader always follows a strong vision of the future and ensures that all changes immediately and continuously contribute to the success of the business. More significantly, a business that truly experiences a transformation in its ability to perform will, over time, successfully evolve and routinely change in significant ways uniquely appropriate to the circumstances.

Case Study: Situations beyond Your Control

In Exxon's synthetic rubber business, we did not anticipate the Gulf War and certainly were not prepared to have our primary raw material quintuple in price. However, because we were able to respond with great agility when that situation arose, it made us more competitive in terms of immediate performance as well as in our long-term sales volumes and customer relations. We used our new lean ability to perform well in an unanticipated way mandated by changes in circumstances beyond our control.

We did the same thing at Suncor. When we began lean, we did not anticipate that the world price for crude oil would drop from $150 to less than $40 per barrel within just a few months. However, when it did, we used our new capabilities to reduce our unit production costs greatly; like Exxon, we performed well in an unexpected situation as required by changes beyond our control. Today, the marginal profit for most oil sand producers is tiny or negative, but ours is quite good.

Sustaining Improvement

When we talk about "sustaining improvement," we are referring to the institutional ability and agility to respond to the constantly evolving needs of the world around your business and the people within your business in a way that continues to produce an ever more competitive business result. *Once you have created it, you need to sustain the ability to improve your business in meaningful ways at a good pace.*

Note: The improvement that needs to be sustained is not the absolute value of whatever performance you have already achieved or the set of operating practices that you adopted to enable your initial improvements. The improvement that needs to be sustained is the ability to achieve even better performance and to adopt ever better practices continuously.

Key idea: This is the industrial equivalent of money management. No matter how much money you have, it will not be satisfactory simply to keep that amount. If your future pace of acquiring new money is less than the rate of inflation, your relative financial position in the future will constantly become worse than it is today. In industry, no matter how good your current performance is, if your pace of improvement is less than that of your competitors, your competitive position will erode.

Transformational change requires that *today* you recognize what is important to your business and respond well to that challenge or opportunity. Sustaining change requires that, *at all times in the future,* you are equally able to recognize what is important and able to respond equally well to those challenges and opportunities.

Process Documentation

There are, of course, managerial aspects to sustaining change in addition to the leadership challenge of creating change. When a single minute exchange of dies (SMED) team demonstrates that it is possible to service a heat exchanger with no lost time, the experience needs to be documented so that future teams achieve at least that result. A number of very good systems exist for documenting this information that will fill management's need to preserve known practices and ensure their use as a basis for future performance. You will want to identify such a system that fits your needs and adopt it. *The system itself is not important so long as you have some formal way to preserve and use specific knowledge as you create it.*

In many situations, the innovative use of documentation systems becomes a good part of your frontline improvement process. At Suncor, we have begun to use work records and instructions that include a great many digital photos of the work in progress, including the tools, lifting devices, and equipment that enable the work. This is not only a good way to record our standard job instructions, but also a great way to make it easy to transfer learning among individuals and teams. Possession of a well-documented standard practice is the basis for all future improvement work; as a manager, you must ensure that this need for sustaining information is well met.

The *leadership* of sustainment, though, is quite different from the *management* of standard practices. Leadership requires that, when a SMED team demonstrates that it is possible to service one heat exchanger with no lost time, that knowledge must be promptly deployed further to all other exchangers and carried forward to everything else that is like an exchanger. Sustaining leadership also continuously assesses the situation to determine what further opportunities exist now that time is no longer lost in servicing exchangers. In addition, sustaining leadership continually assesses the entire operation to determine if an unobvious improvement in some other aspect of the business might be more valuable than further work related to exchangers.

For example, at Suncor, by assessing the entire business, we first used the increased capacity we gained in extraction to increase the intensity of sand removal in a way that would reduce erosion in the upgrader rather than to increase production in extraction. This alternate use of new capacity in one place to improve operations in another place was the best outcome for the business but it was only possible as a result of taking a businesswide view. Continuously improving and continuously extending the scope and value of improvements, often in unobvious ways such as this, or identifying and capturing the opportunity hidden within disruptions such as the Gulf War or the precipitous decline in world energy price is a clear signal of effective sustaining leadership.

THE ROLE OF TRANSFORMATIONAL LEADERSHIP

Thus far, our emphasis has been on engaging people at the frontline of the business. Virtually everyone in a typical Western manufacturing enterprise needs far more help from the frontline than we receive. Lean is a wonderful path to engage the frontline teams. Every aspect of lean is specifically designed to work well in a highly engaged team structure and every aspect of lean practice encourages you as well as those on the frontline to engage with each other to make it happen.

In this chapter, however, our focus is on leadership. Creating and sustaining lean or any significant systemic change that transforms an enterprise is a matter for leaders. In Western industry, the vast majority of people are good, honest, hardworking people who are constantly making their personal-best contributions to the enterprise. However, those contributions are strictly defined by the enterprise management system and necessarily conform to the expectations and constraints of the system. In a traditional industrial management system, the best contribution available to many people is simply executing a specified task. In a more engaged system such as lean, people can make more meaningful contributions, but most people's contributions remain limited in a way that is defined by the system.

The system within which people work belongs exclusively to management. Interestingly, most managers and executives are working on different tasks and at different organizational levels than the people in frontline teams. Like those in the frontlines, the vast majority of executives are honest, hardworking people who conform to the existing system. They are fully competent at complex, high-level tasks, but they are not leaders. *Transforming the system within which people work always requires a leader.*

When a leader creates a culture of improvement and transforming change, the executives, managers, and frontline teams will quickly learn how to perform competently in accordance with that new system. Once people conform to a new system of continuous improvement, both the pace and the magnitude of improvement realized within the business will far exceed what any individual could accomplish alone. *Leadership is not a solitary effort and the effects of leadership always properly appear as the work of the larger team. However, without at least one leader who creates the transforming change to the system within which the team operates, no systemic change will occur and an unexpected magnitude of performance improvement will not be achieved.*

One of the principal attributes of a transformational leader is the ability to envision what the transformed enterprise will be and how to achieve that vision. As you get started with lean, you should develop a clear vision of what it will look like in your business as it matures. Then, communicate your vision to your colleagues and let them help you shape it and achieve it. The gap between where you are now and where you envision your business to be in the future will define the earliest steps on your path forward. Early communication of your shared vision is critically important. Mutual recognition that your business is making successful initial progress toward a shared vision will give you and your colleagues confidence to do even more as the transformation progresses.

The second attribute of leadership for these purposes is the ability to attract people to join in the pursuit of the shared vision of the future. In most industrial settings, this is not a matter of inspirational leadership and persuasion as much as it is a matter of pragmatic leadership by demonstration and education. Lean theories and practices provide a leader with the ability to implement and demonstrate transforming change that will attract the attention of people throughout the enterprise.

Once that occurs, others will want to do what you are doing. At that point, educating others to enable them to do what you have done will initiate the process of spreading lean throughout your business. In addition to the confidence that your team has gained from the successful pursuit of your initial objectives, deploying lean through example and education provides your colleagues with the further confidence that lean can be learned, taught, and practiced with a predictable outcome. *Providing people with confidence in both outcome and practice is an important element of engaging others to share lean with you.*

Stories or urban legends abound about inspirational leaders who, in special circumstances, transform organizations to new levels of performance. However, transformations by inspiration often occur in ways that others cannot sustain and that cannot be reproduced elsewhere by anyone, including the inspirational leader who tries to do it again in another place. Lean manufacturing is not like that. Most people can practice lean successfully and the results can be reproduced as often as you want. Rather than benefiting from lucky circumstances on the first try as inspirational leaders often do, lean leaders frequently report that, like most technologies, lean practice becomes easier and more successful with experience. As you commence lean implementation, make sure that you do so based on well-documented theory and practice. Help people throughout your business create their own examples and experience of using that practice in their own work.

Key idea: Lean does not necessarily have to be initiated by the CEO or by any other C-level executive. It is entirely possible for a person in any managerial position to commence transformation of the portion of the enterprise that he or she leads by using lean methods within his or her existing span of control. If leaders are able to add education to their examples so that other managers can understand how to replicate their success in the portions of the enterprise that they lead, the transformation will spread like a virus. Every example in this book is derived from my experience, and I have never led the entirety of an enterprise.

A very interesting aspect of leadership is that, as you commence transformation of a business, natural leaders will spring up throughout the organization, each one ready to lead his or her part of the change. Each of these leaders will make a valuable contribution to establishing the new system in the detail appropriate to each particular part of the operation. Some leaders will be existing managers; others will not. Often the informal leaders add a value to your effort that cannot be reproduced in any other manner. Informal leaders at the frontline often have personal credibility among their peers that no manager can ever achieve. Among the most valuable contributions of any transformational leader is the creation of additional formal and informal leaders throughout the enterprise who can serve as "disciples" to spread the change.

Giving each of these disciples an independent capability to lead in the details appropriate to his or her work is important. During the transformation at Gilbarco described in Chapter 1, I purchased several hundred copies of Richard Schonberger's book, *Japanese Manufacturing Techniques.* At Exxon Chemical, we widely deployed a series of 11 white papers that I wrote describing my early understanding of lean practice in the chemical industry. During the transformation at Suncor, we deployed several hundred copies of my earlier book, *A Culture of Rapid Improvement.* In each case, giving this information to many people enabled them to develop the confidence that comes from possessing independent knowledge of the technical processes that we were following, and it allowed people throughout the business rapidly to develop personal expertise that enabled them to lead their part of the transformation.

More importantly, by widely distributing the formal version of the path we shared, it became possible to accelerate the pace of implementation with much greater assurance that we were all doing the same right things in the same right ways. I recommend that course of action to you as well. If you want people to implement lean by the book, give them a copy. Holding the source of your path forward to yourself will not engage enough people to help you succeed. The most important attribute of industrial leaders is not that they are the sole source of wisdom and knowledge. *The only important attribute of leaders is that they have successful followers.*

Sustaining Leadership

In precisely the same way that a leader is required to initiate a transformative change, a leader is required for the task of identifying and directing sustaining change. The leadership challenge is essentially the same. Someone who has the credibility and the capability to change the system within which others work needs to chart the course for the future continuously and ensure that the organization has the ability, agility, and discipline to follow it.

Your business, your people, your competitors, the global economy, and the communities in which you operate are all in a constant state of evolution. There is no possibility that you will ever succeed in sustaining your improvements by establishing a rigid course and pursuing it inflexibly. That would be the industrial equivalent of steering a ship by looking at the wake to ensure that you are always going in a straight line. Traveling in a constant straight line may be a highly disciplined way to operate a ship, but it is a useless way to reach a complex destination, especially one that is certain to have obstacles along the path. Whether the ongoing change required is large or small, great businesses always depend on the good advice of a person who can initiate and successfully lead ongoing change.

Key idea: *Transformational leadership and sustaining leadership* are different, but not very different. They *are conceptually two sides of the same leadership coin.*

When the Leader Is Not the CEO

Lean does not need to be initiated or led by the chief executive officer or by any other C-level executive. That is a good thing because the CEO normally has many other things to do. However, senior executives do need to *understand* and *support* or enable the transformation. If you are leading the change, you will need to find a way to communicate upward in addition to communicating with your colleagues.

Key idea: Senior executives do not need to lead the transformation in order for it to begin and advance. However, a senior executive who does not understand the nature of the big, fast change that is occurring within his or her organization can easily stop it even without intending to do so.

Senior executives have two great opportunities to engage with the transformation in the *wrong* manner and they both occur after the transformation has begun but well before it is complete. The first arises because the new performance is achieved rapidly and, in the nature of leadership, the new performance is achieved through the actions of the larger team. Senior executives who do not know objectively what was done to achieve this outcome often conclude from the speed and ubiquity of the result that the new performance was just about to occur in any event as a result of other causes.

The other potential for inappropriate executive interaction is that the magnitude of the early performance improvement is very impressive. Thus, senior executives who do not have a clear understanding of the cause and the potential of this transformation make the mistake President G. W. Bush made and declare "mission accomplished" when the race is really just starting.

Either or both of these mistakes can easily and inadvertently result in the same adverse outcome. If senior executives believe that the new practices are not linked to the new performance or if they conclude prematurely that the new performance is so good that the transformation is complete, they can diminish or deemphasize the transformation while it is immature; the result is that it may never reach its full potential. Lean operations cause improvement by transforming *every* job in the business into a more intense, more focused, and more engaged job. Similarly to other new activities, initially this new experience is attractive and interesting and, if pursued properly, in the long term it can become customary. However, there is an interim point where the increased intensity is no longer new and the extra work is not yet customary.

As with all transformational changes, this is a "tipping point." When it is obvious to everyone in the organization that both a "new way" and an "old way" exist, people begin to watch the behavior of senior managers very carefully. At that point, in the minds of the observers, senior management must either embrace the change or support the heritage. If senior managers ignore the change and continue to deploy the visually apparent tools of the organization—recognition, reward, and inclusion—in the same ways that they always have done, then the extra efforts are deemed to be unsupported and the transformation will not pass the tipping point. The potential to achieve world-class performance will become simply another interesting artifact in the corporate history. The immense gains that are possible will never be realized, and no basis will exist for sustaining the competitive advantage of the gains that have already been achieved.

Key idea: Lean technologies always have the capability to transform the performance of any manufacturing organization. However, corporate transformations sometimes do not begin well because they are undertaken by a traditional executive as a *management project* rather than by leader as a *transformation.* Lack of leadership is the cause of "copy cat" implementations that frequently become so bogged down with administration that great new practices are used in surprisingly bad ways. Sometimes, transformations begin well and then fail to achieve their full potential because transformational leaders do not successfully engage senior executives.

GETTING STARTED

As Stephen Covey writes in *The Seven Habits of Highly Effective People,* we need to "begin with the end in mind." As you commence adoption of lean practices, you should do so intending to create an immediate benefit within the first 6 months. Thereafter, you should anticipate an ongoing and accelerating pace of improvement that will sustain your competitive advantage by creating still further change indefinitely.

Modest initial results, delayed benefit, or an organization that ceases to improve after beginning well are all indications that your lean practice is not being conducted as it should be. Lean is indeed a *technology* of manufacturing practice. As with other technologies, it can be learned and practiced and the outcome is predictable. More importantly for many people, if it is not initially successful, it is a practice that can be assessed and improved. By focusing on the demonstrated performance and the ongoing pace of your transformation—both measured in a continuing series of relatively short time periods—you can ensure that you get on track and stay on track to the performance that you desire.

The Value of 6-Month Intervals

I always recommend that a transformation be planned and executed in a series of 6-month intervals. Six months is neither too long nor too short. In a period shorter than 6 months, you risk the possibility that other things may distract you from achieving measurable progress. You have a business to run, and that often makes immediate operating demands that you and your team must meet. Further, a period shorter than 6 months may not be sufficient to be certain that you are measuring performance that resulted from intentional changes rather than from natural variation. Even faced with other demands, over a period of 6 months you should have time to achieve an objectively observable change in performance that is easily distinguishable from normal variation.

A period longer than 6 months reduces the urgency with which the team approaches the work of transformation. By a wide margin, the most common mistake that people make when starting a new way of leading manufacturing is that they simply fail to get started. This effect often manifests itself in long periods of planning, training, developing new systems, and other such things. When there is a long period before results are expected, the implementation team becomes comfortable spending its time preparing for the future rather than implementing an immediate change. A period longer than 6 months between measurements of progress makes it likely that less progress will occur.

The first rule of getting started is that you should plan, execute, and measure your implementation in relatively short periods. In capital projects, the mode of change is to make an investment and experience the benefit after the investment is fully implemented. In continuous improvement, the mode of change is to commence the transformation immediately in accordance with the shared vision and experience the initial benefit of those changes within 6 months or less.

Three Attributes of a Successful Beginning

As we have discussed, three attributes always surround a successful leader:

1. A leader has a strategic vision of the future and shares that vision with others in a way that makes the shared vision meaningful to each person or team.
2. A leader engages people in a way that provides each individual or team with a personally relevant opportunity to make unique contributions to the successful attainment of the shared vision.
3. A leader provides people with new capabilities or tools that make it practical for them to make a new contribution to a new vision in a new way.

These three attributes of success—goals, engagement, and tools—are a part of every step in your lean practice and they represent a critical part of how you get started. As we proceed, we will use these elements to describe a path forward that you can adapt to fit your organization, business, and people. This description will focus on the first 6 months. Once you have finished your first 6-month period, you should be well on your way. I have also included a summary description of the objectives that you should plan to experience during the second, third, and fourth 6-month periods of your implementation.[1]

Recognizing that the most common mistake in getting started is failure to start, you should plan the first 6 months of your implementation with the goal that, in addition to laying the foundation for the future, you will absolutely produce objectively measurable improvement within that time. With that in mind, here are examples of the three elements of leadership, as you may want to practice them during the first 6 months.

The Value of Shared Vision

A formal practice of goal setting and goal translation or policy deployment is an ongoing essential component of an established lean enterprise. During the first 6 months, you should begin that formal process of goal development and policy deployment with the intent to have well-developed and well-communicated goals deployed throughout the organization by the end of the period. You will experience, as we did at Exxon and Suncor, that goal deployment has inherent immediate value and that you will get better at doing it in a continually more valuable way as you repeat it over time.

Key idea: You should begin your improvement practice by promptly providing the people who will help you get started with a meaningful vision of the immediate future commencing at the very beginning of the first 6-month period.

The Value of Immediate Pilot Projects

At the same time that you are formally looking several years into the future and describing your strategic objectives for the entire business over a long period, you should begin immediate improvement by looking only 6 months into the future and describing initial objectives that can be realized within that more limited period. For small businesses, you may describe a limited but immediate objective applicable to the whole business. For larger businesses, you will normally start by describing limited immediate objectives for one or more subsets of the business where you intend to begin teaching people to use lean through pilot projects that you can initiate immediately.

In the future, as people fully engage, they will want and need a longer strategic horizon to ensure that they work autonomously in ways that are assuredly additive and compatible over long periods and across the entire organization. But you will not start your improvement effort with autonomous work and, unless your business is very small, you are not likely to start with the entire organization. The need for complete and more rigorous goal deployment throughout the enterprise can be successfully met at the end of the first 6 months.

The interim goals that you use to initiate pilot improvements immediately do not have to be communicated generally or even formally. Deploying goals to people who are not yet participating in the pilot projects or with whom you will have a close relationship as you assist them to implement the pilot projects has little value. If you delay the start of the transformation in order to provide a complete and formal goal set to everyone, when a lesser goal set delivered informally to a few people will serve the immediate needs very well, you will simply slow your overall rate of progress. Your progress will be slowed because your business will not have the benefit of your pilot projects and because your ultimate goal deployment will not have the support of completed demonstrations of the new capabilities. *You should begin lean implementation by selecting an immediate goal and deploying it only to the people who will help you achieve it and then using that achievement as an exemplary demonstration of the new practices.*

Case Study: Start Small

As we began lean implementation at Suncor, unreliable equipment limited our operations. Several months before we began formal deployment of a complete goal set throughout the broader enterprise, we established the immediate and more focused goal that, within 6 months, a few key operations would no longer be limited by reliability problems. By establishing and achieving that obviously challenging but quite focused goal, we provided both an immediate benefit to the enterprise and credibility to the improvement effort.

As part of achieving the goal of immediately improving reliability at Suncor, we conducted our improvement activities with great focus. *We worked on things of real importance and did not work on other things.* For example, we had a long-standing "improvement team" in one area that had achieved very little during the prior 2 years. In fact, that "team" consisted of one-half of one person. Because that was a critical area for improvement, we promptly redeployed 11 people to that team and achieved the intended improvement within a few months.

We began with reliability in our upgrading operation because it is the last operation before the product enters the pipeline for transport to our customers. By starting with the final process, we practiced what is known as a "clearing" strategy. After our improvements to the upgrader were complete, all of the product produced in our extraction unit could clear the plant through the upgrader.

Continuing in that clearing sequence, we moved next to the extraction process and improved that unit until it could process everything produced by the mine. Improving equipment from the back of the plant toward the front so that each unit can clear all the material that reaches it is a classic strategy to improve the capacity of a generally constrained plant that lacks a well-defined bottleneck. It is also functionally identical to the lean theory of "demand-pull production."

Note: I have not previously mentioned the concept of "demand-pull production" because it is most appropriate in discrete manufacturing and rarely appropriate in chemical manufacturing. The concept is based on the idea that customer or market demand "pulls" product from your last operation, which in turn creates an internal demand that "pulls" product from prior operations. In this sequence, a sufficiently flexible, reliable, and capable plant can essentially terminate short-term production planning and simply link operations directly to real-time customer demand. If this concept is appropriate to your operation, it is well documented in the literature.

As you commence lean, you should find your own version of our reliability improvement initiative (immediate and focused objective) and promptly begin there. Every plant has one or more apparent, intractable problems that are highly visible and where the resolution is obviously valuable. Resolving one of these obvious problems will certainly align with your ultimate goals. Find one that is amenable to improvement according to the lean tools or by any other improvement tool that you can rapidly deploy. With a limited scope and with a limited tool set appropriate to the task, get started by setting and achieving the goal of immediately solving that well-known problem in an exemplary manner.

In addition to communicating the goal of resolving the immediate opportunity for improvement, you should communicate the other elements of transformational leadership. Introduce people to the tools that will be used and the expectation of enhanced engagement with people at the frontline. Treat each new activity during the first 6 months as a pilot project that visibly demonstrates the new way of improving the business. Every successful example should be part of the learning experience for the organization. You want people to learn about shared goals, engagement in improvement, and new tools of creating improvement. You also want people to learn that you are a credible leader of change and, of course, you will want to support your initial pilot projects in a way that ensures that they are both exemplary and successful.

Thus, the first step on the road to lean is to undertake one or more pilot projects focused on an important and highly visible problem. To begin, immediately communicate the vision or goal of resolving that problem by engaging people in the use of the new tools of lean manufacturing. In this way, you can

- solve a problem that delivers immediate benefit
- teach people how to use a new tool
- engage people in the use of the new tool
- give the organization confidence that you are leading it toward improved capability to do things that were not previously possible

For these initial pilot projects, you should always pick a big, important problem. Teaching people that lean can solve easy problems that could be resolved in other ways has very little value. In addition, deploying lean toward a project that has form but no substance has no value. You can confidently start with an intractable problem because, as a management-led and -supported pilot project, it will benefit from the focused attention of your best implementation team in a way that future projects will not.

You can and should deploy as many pilot projects as you believe you and your team can successfully manage to completion. In this way, within a short period, you will not only have your own examples of how lean is practiced in your business, but also have a nice body of improvement.

Case Study: Reality Check

Lean is a technology set for improved manufacturing—not a panacea. Exxon once owned a coal mining company that produced coal with a very low energy content. A large quantity of this coal produced only a modest amount of heat. As a result, it was inappropriate for many purposes that required intense heat; therefore, its price generally was low, which made it unprofitable to produce. As we began to experience increasing success with lean manufacturing at Exxon Chemical, the coal company executives frequently requested that we help them use lean to improve their mines. Unfortunately, no aspect of lean technology can increase the energy content of coal. Exxon later sold those mines and exited the North American coal business.

It is perfectly acceptable to set very aggressive improvement targets. I encourage you to do so, but make certain that the targets you set are appropriate to the technology that you are deploying.

The Value of New Tools

Progress always starts with a vision of the future. Quite literally, you cannot possibly be a leader if you do not know where you are going in sufficient detail to describe how your staff can help you get there. In most instances, immediately after you communicate your goals, you will recognize that if people were currently able to perform at the level you envision, at least some of them would already be doing it. Achieving a new level of performance almost always requires that you provide people with new capability or with a demonstrably different and better way to utilize existing capabilities. Everyone in industry has already experienced managers who simply set challenging goals and do nothing beyond exhorting people to meet the new goals. *There is no value at all in aspiring to obtain a new result while continuing the old way of doing things, and everyone knows that.*

During the first 6 months, as you develop the complete set of goals that will eventually be deployed within your business, you will also want to identify the tool set that is appropriate to your plant, your people, and your goals for improvement. For example, if you have essentially a single product operation as the Suncor Oil Sands Group does, you may decide that you have no interest in the FSVV (fixed sequence variable volume) tool. If you have a small batch process, you may decide that you want to know more about demand-pull scheduling. Most plants will want some form of SMED, operator care, and mistake proofing and, as your improvement effort matures, you are likely to engage in statistical process control. The key message is to make the selection of the tools to be used and the timing for adopting each one a formal part of your management plan for lean implementation.

As described in earlier chapters, there is a formal practice for developing and deploying each of the several new tools of lean throughout the organization. For each tool in its turn, you will want to develop an experienced subject matter expert with the capability to provide the tool to your teams and support their successful use of the new technology. A principal way to develop that capability within your organization is to deploy the tool in the early pilot projects where you can train people who will immediately use the tool in your plant. In general, lean tools are often quite intuitive, but they are still new tools and merit formal training as well as expert mentoring when put into widespread practice.

During the first 6 months, you should not devote any significant effort to mass training. That is a classic trap; it will consume your time and money with no resulting benefit. As you deploy new tools, you must avoid treating this as a training exercise. Your business objective is to create people who *use* new tools, rather than people who merely *possess* new tools. As you begin, teaching everyone about all the tools has no immediate or long-term value. Start deploying new tools by teaching those who will use them in your pilot projects and those who will later become your subject matter experts.

Key idea: Start by teaching a few people the few new tools that are directly associated with the immediate objectives that you have set for completion within the first 6 months.

Both your subject matter experts and the few others who are engaged in the early pilots will learn the new tools and sharpen their skills immediately while conducting the initial projects. The team members for your first pilot projects will naturally be among your best people; this is how participants in important pilot projects are selected. In addition, the projects will be closely led to ensure that they are successful and exemplary. These pilot projects will be great learning experiences. Management will assure this outcome by providing the attention, resources, and assistance that early pilot projects deserve. These special people will bring their special capabilities to the difficult tasks selected for your initial pilot projects and, with proper support and development, they will later serve you broadly as subject matter experts or team leaders to teach and help others to use the new tools.

Key idea: Always train people principally in the tools that are appropriate to their improvement goals. In that way, training will always produce a nice return on the investment. Providing formal training only for the new skills appropriate to the formal improvement goals is as applicable in the first 6 months as it is throughout the course of lean practice. If people or teams need other capabilities as their practice matures, they can receive additional training at that time, or those capabilities can be supplied on an ad hoc basis by your subject matter experts.

Learning about and using new tools are the fun part of lean for most people. The tools of lean (see Chapters 4 through 9) are all very powerful and they are all very useful for frontline teams to achieve things that were not previously possible. Most people instantly recognize that this is a new way of manufacturing that will result in real change. Those who become your disciple leaders will immediately want to join the fun.

Case Study: Truly Powerful Tools Attract Users

At Suncor, when we initially taught the SMED technology, we began with one 2-day course for 20 people. The morning that the first course started, only 18 people showed up and I was making phone calls at 6:30 a.m. to find the stragglers. By the morning of the second day, the word had spread that this training was indeed an important contribution to the way we work. Participants were describing SMED as the best industrial training they had ever received. People who were not part of the first course began asking when they could get a chance to participate. All our classes since the first class have been oversubscribed and well attended.

Lean has many elements, but the one that always attracts the most immediate and favorable attention from veteran manufacturing leaders is the bundle of new tools for use at the frontline. It is always surprising how many people voluntarily find an immediate way to practice a tool that is especially appropriate to their work and within their existing authority. (This is another benefit of providing your staff with the book that you are using to implement lean.)

For many people, including me, learning about a great new capability immediately invites an attempt to experience it in their own work. This individual experimentation is not a part of a formal pilot project and it is not autonomous improvement (because it is within existing authority and not part of a new, well-managed system of autonomy); however, it is always a great benefit to the overall progress of the organization. I have seen electricians rewire panels to make the state of the operation visually apparent in the manner of mistake proofing. I have also seen mechanics adopt SMED to remove waste from their work and supervisors use the principles of 5S to arrange their work teams and work areas for improved work flow.

Most lean practices do not require formal management of change and most are well within the existing authority of many people. If that is the case, let people experiment a little within their existing limits, not only during the first 6 months but also throughout the implementation period. People can engage with lean prior to establishment of a formal practice of autonomous improvement so long as they do it within their properly existing authority. There is no reason to restrict innovative actions that otherwise would be appropriate.

Within the formal management-sponsored pilot projects that are intended to create an exemplary change, the selection of, formal training in, and use of the new methods must be more rigorous. In that process, you are developing your future subject matter experts and they need true expertise. However, similarly to goal deployment, the immediate need that arises during the pilot projects is not to deploy the entire lean tool set formally, but rather to deploy only the particular lean tools that are likely to contribute to the immediate goals that you have selected to begin your implementation. In order to get off to a good start, not everyone needs all the tools. A few people need the *appropriate* tools required to implement the first successful examples of how lean is practiced in your business.

If you select an initial goal for a pilot project that is obviously amenable to resolution with a lean tool, then the first tool to roll out is clear. No specific pattern tells you which tools are appropriate for your first use. That is guided by the project you select. In my own practice, I have generally started with operator care, SMED, or FSVV because the easiest problems to recognize and resolve in a new way are often associated with reliability or flexibility. However, it really does not matter where you choose to start. The immediate issue is to achieve prompt improvement to the business and to demonstrate that you are leading people to a new result in a new way.

Key idea: As you proceed through the first three or four 6-month periods, make sure that you have at least one pilot project that will demonstrate and teach each of the lean tools that will ultimately be a part of your tool kit. This will be the only time in your lean experience when the tool will drive the project selection.

Case Study: An Exception to the Rule

The very first tool we deployed at Suncor was not a lean tool. Just prior to initiating lean, Suncor experienced a poor implementation of the SAP enterprise computing tool. Therefore, the first thing that we did as we began to pursue our goal of improved reliability was to begin using SAP in a new way that resulted in people, plans, and parts coming together at the maintenance job site for the first time. That was not a lean tool, but it was obviously a new way to use an existing tool, which is often an equally valuable contribution to immediate improvement. Indeed, the fact that we preserved, enhanced, and incorporated an existing capability, instead of abandoning it, was a valuable part of establishing the idea that lean is a continuous improvement effort rather than just a new program that supplants older programs.

Many tools are available for improved manufacturing and lean offers some great ones. However, the critical issue for leaders is to provide teams with a new capability that will give them confidence to attempt and achieve a result that previously eluded their best efforts. As you practice lean manufacturing, some of your people should become masters of each of the lean tools. They are at the heart of a system of manufacturing that fits together seamlessly. However, you should also feel free to use other tools as appropriate to your circumstances. Good lean practice is enabling, not limiting.

Early on, as you expose your team to the theory and practice of lean manufacturing, along with formal pilot projects, you should anticipate that "skunk works" will arise spontaneously within the personal authority of individuals throughout the organization. The formal pilot projects must be carefully managed and supported to succeed in an exemplary manner. During formal pilot projects, the new tools of lean need to be taught carefully as you develop formal experts for the corporation.

You will also want to watch the informal projects closely. Do not allow anyone to do something for which he or she has no existing authority. Do not allow anyone to use the "words of lean" to do things that are not lean at all or to use the lean concepts or tools in an inappropriate manner, such as reducing inventories of spare parts prior to improving the state of maintenance so that the parts are no longer required. Lean is not a license to violate your policies for the management of change, and lean practice becomes authorized for autonomous activity only when all the attributes of carefully managed autonomy have been created.

Beyond those cautions, most people who have authority to try new things will do so when they are reasonably confident that they will do it in the right way and achieve some success. Some people will surely surprise you in the ways that they succeed when given a new and attractive tool. Take advantage of that. You may also want to provide at least some support to ensure that most of these early volunteer efforts are obviously successful and exemplary.

The Value of Engaging People in a New Way

We have been discussing the engagement of people, including creating an autonomous system within which people can assuredly do the right things in the right ways with careful management but little supervision. The only possible way to achieve a world-class pace of improvement is to allow people to make improvements with some degree of autonomy.

However, autonomous improvement is among the last things that you will introduce as you practice lean manufacturing. Certainly, there should be no expectation of new autonomous work within the first 6 months. Autonomous work is always carefully conducted within a well-constructed system (see Chapter 10). *Within existing authorities,* in the first 6 months you can allow informal pilot projects using lean technology, but there can never be autonomous work that exceeds the existing formal authority of either individuals or teams.

Key idea: The most common outcome of unauthorized or premature autonomous action is that some people commence doing the wrong things in the wrong way. That will invite management intervention in a way that has a long-lasting adverse impact on your future efforts to initiate proper autonomous activities. It often seems like a good idea to let people loose to pursue their individual ideas of improvement, but you are always well advised to limit that practice strictly until it can be assuredly successful as part of a well-managed formal effort.

However, you can immediately begin engaging people in a new way of working as you conduct the pilot projects by giving them the goals for the pilot projects in their areas; teaching them to use the new tools, which will be part of resolving the opportunities represented in the pilot projects; and allowing them to begin to practice working with those tools with management support and assistance. This will not be autonomous action because the pilot projects will be closely managed and the engaged people will be practicing the new tools with formal assistance and under close supervision.

Key idea: Autonomous improvement is a special, more advanced form of engaging people to help you improve. You will prepare people for later autonomous work by teaching them to use new tools in new ways as part of carefully managed pilot projects.

You will be seeking three benefits as you engage people in the pilot projects. First, you want to use and demonstrate the use of the new tools of lean. Those tools were specifically designed for people and teams at the frontline. As you deploy the tools, you will want people at the frontline to engage with you in their use. You will be using the pilot projects to implement change but you will also be using the pilot projects to teach people the new tools and to develop your subject matter experts formally.

Second, you will want to teach people at the frontline to practice small event improvement. This is very different from the big event improvement projects that they will have experienced previously. Obviously, your pilot projects will be well supported and closely managed, but it should always be obvious that the people at the frontline are conducting the project using the new tools of lean or other new capabilities. It should be equally obvious from the conduct of the pilot projects that similar projects could be successfully conducted autonomously in the future within the capabilities and resources of the frontline team.

Finally, you will want to demonstrate through this initial engagement of your frontline teams that improvement practiced by frontline people using new tools results in goal-focused progress that makes a meaningful contribution to the business.

At the end of the first 6-month period, you should at least have

- created a formal, written, long-term business strategy that you have deployed or promptly can deploy throughout the organization
- identified the initial set of new improvement tools that you will use in your implementation and begun to develop subject matter experts on each tool
- identified and implemented one or more management-supported pilot projects that provide an immediate business benefit in an exemplary manner
- engaged people as part of the pilot projects in a new way, using the new tools in pursuit of the new goals and in a manner that obviously can lead to future autonomous work
- allowed people or teams who have existing authority to experiment informally with the new methods

At the end of the second 6-month period, you should at least have

- utilized the corporate strategic goals that you have deployed to create very focused goals for each frontline team that are identically aligned with the corporate goals that you have developed
- expanded your pilot projects to more areas of the plant—each one using the goals of the area to select the appropriate opportunities for improvement, thus creating immediate new value with each pilot project
- expanded your tool set to include tools appropriate to the new set of pilot projects
- expanded your cadre of subject matter experts through additional training and through experience with additional pilot projects
- engaged more people in the additional pilot projects
- begun to define the formal steps that will create and manage your developing system of autonomous improvement

At the end of the third 6-month period, you should at least have

- engaged people at the frontline with quality stations and a fully developed practice of autonomous improvement
- developed a capability to train people as needed or to provide ad hoc assistance to enable all teams to use all the tools in your tool set, as appropriate, to their developing goals
- created a formal and disciplined practice of objectively qualifying people to be improvement team members, team leaders, and subject matter experts (separate qualification for each tool) and to qualify those who will mentor your team leaders and subect matter experts
- conducted pilot projects of immediate value throughout the business, still closely supervised and supported but clearly demonstrating the autonomy that is developing

At the end of the fourth 6-month period, you should have fully functional lean practices evident throughout your enterprise. Throughout this period, you will

- continue to expand on the success of the first three periods
- conduct further goal-focused pilot projects
- engage more people and teams at the frontline, including authorizing autonomous action for the teams who are ready
- deploy more tools and develop subject matter experts for those new tools

Your lean implementation will continue to evolve and mature for several years, but it should be obvious in 2 years or less that you have transformed your organization in both form and substance. You should start your first 6 months with the goal of achieving that.

Getting started is not really so difficult. It is generally just a small and fast version of promptly using the long-term capabilities described throughout this book and continuously building on that initial base. You should begin by immediately doing something that is of real value to the business in the first 6-month period. During each 6-month period thereafter, you should make demonstrable progress on the three elements of leadership: goals, tools, and engagement. During each 6-month period, you should make good progress toward full deployment of all the capabilities described here that are appropriate to your business and your people. In addition, during each 6-month period, it should always be clear that the business is objectively better because of this work.

The performance described here is not theoretical; it is all very real and it is all from my experience. These capabilities have allowed us to take good businesses and transform them into great businesses. They have allowed us to take struggling businesses and make them very good businesses that are on the road to becoming great businesses. They have allowed us to prosper in the face of disruptive change, and they have allowed us to create disruptive change that benefited our businesses in a unique manner.

Whatever your situation is as you commence this work, when you adopt these theories and practices in your business, you too will be on your way to world-class performance.

NOTES

1. If you want an even more detailed description of the first steps in leading a transforming change, four chapters in my earlier book (*A Culture of Rapid Improvement: Creating and Sustaining an Engaged Workforce,* CRC Press, 2008) are devoted to the process of getting off to a good start.

Index

L

N

About the Author

Raymond C. Floyd is senior vice president of Suncor Energy. Prior to joining Suncor, Ray retired from Exxon Mobil, where he spent more than 20 years and where he most recently served as global manager of manufacturing services. Previously, he was with General Motors for more than 10 years.

Ray is generally recognized as one of North America's "early adopters" of lean manufacturing and is among the very first worldwide to adapt lean technologies for use in the chemical and process industries. Following the practices described in this book, Ray led the first chemical business to receive the Shingo Prize and has led two separate businesses that have been designated as one of "America's ten best plants" by *IndustryWeek* magazine. Ray is the only person to lead businesses in both chemical and mechanical manufacturing to receive that designation. As site manager for Exxon's massive Baytown chemical plant, Ray led the team that was designated as "best maintenance organization in large industry" by *Maintenance Technology* magazine. Ray received the Andersen Consulting award for "excellence in managing the human side of change."

Ray has degrees in chemical engineering, business administration, and law. He is professionally licensed as an engineer, attorney-at-law, and patent attorney. He has also received international senior executive development at the Institute for International Studies and Training in Japan and the Institute for Management Development in Switzerland. Ray was appointed by President Reagan to represent the United States at the Japan Business Study Program as a guest of Japan's Ministry of International Trade and Industry.

Ray's wife, Marsha, is also an attorney-at-law. Ray and Marsha have two daughters, who are both physicians, and five grandchildren.